STUDENT SOLUTIONS MANUAL

to accompany

CALCULUS

SINGLE VARIABLE   SECOND EDITION

Deborah Hughes-Hallett
Harvard University

Andrew M. Gleason
Harvard University

et al.

John Wiley & Sons, Inc.
New York • Chichester • Weinheim • Brisbane • Singapore • Toronto

This project was supported, in part,
by the

National Science Foundation

Opinions expressed are those of the authors
and not necessarily those of the Foundation

Grant No. DUE-9352905

ISBN 0-471-24260-8

Printed in the United States of America

10 9 8 7 6 5 4 3 2 1

Printed and bound by Bradford & Bigelow, Inc.

CONTENTS

CHAPTER ONE

Solutions for Section 1.1

1. **(I)** The first graph does not match any of the given stories. In this picture, the person keeps going away from home, but his speed decreases as time passes. So a story for this might be: *I started walking to school at a good pace, but since I stayed up all night studying calculus, I got more and more tired the farther I walked.*

   (II) This graph matches (b), the flat tire story. Note the long period of time during which the distance from home did not change (the horizontal part).

   (III) This one matches (c), in which the person started calmly but sped up.

   (IV) This one is (a), in which the person forgot her books and had to return home.

5. For some constant k, we have $F = \dfrac{k}{d^2}$.

9. A possible graph is shown below:

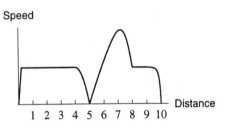

13. (a) What does the statement $f(30) = 10$ mean in terms of temperature? Include units for 30 and for 10 in your answer.

    (b) Explain what the vertical intercept, a, and the horizontal intercept, b, represent in terms of temperature of the object and time outside.

    (a) $f(30) = 10$ means that the value of f at $t = 30$ was 10. In other words, the temperature at time $t = 30$ minutes was $10°$C. So, 30 minutes after the object was placed outside, it had cooled to $10\,°$C.

    (b) The intercept a measures the value of $f(t)$ when $t = 0$. In other words, when the object was initially put outside, it had a temperature of $a°$C. The intercept b measures the value of t when $f(t) = 0$. In other words, at time b the object's temperature is $0\,°$C.

17. Since the function goes from $x = 0$ to $x = 5$ and between $y = 0$ and $y = 4$, the domain is $0 \le x \le 5$ and the range is $0 \le y \le 4$.

21. The domain is all x-values, as the denominator is never zero. The range is $0 < y \le \dfrac{1}{2}$.

Solutions for Section 1.2

1. Rewriting the equation as
$$y = -\frac{12}{7}x + \frac{2}{7}$$
shows that the line has slope $-12/7$ and vertical intercept $2/7$.

5. The slope is $(3 - 2)/(2 - 0) = 1/2$. So the equation of the line is $y = (1/2)x + 2$.

9. The line parallel to $y = mx + c$ also has slope m, so its equation is
$$y = m(x - a) + b.$$
The line perpendicular to $y = mx + c$ has slope $-1/m$, so its equation will be
$$y = -\frac{1}{m}(x - a) + b.$$

13. Since W is a linear function of R, we can find its slope, m, using the formula $m = \Delta W/\Delta R$. Substituting in the first two values in the table gives:

$$m = \frac{\Delta W}{\Delta R} = \frac{25 - 20}{9 - 6} = \frac{5}{3}.$$

Using the first point in the table, we have

$$W = \frac{5}{3}R + b$$
$$20 = \frac{5}{3}(6) + b$$
$$b = 10.$$

Thus, we have

$$W = \frac{5}{3}R + 10.$$

Note that we only needed two of the given points in the table to find W as a function of R.

17. (a) Given the two points $(0, 32)$ and $(100, 212)$, and assuming the graph in Figure 1.1 is a line,

$$\text{Slope} = \frac{212 - 32}{100} = \frac{180}{100} = 1.8.$$

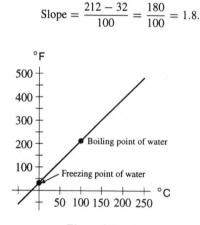

Figure 1.1

(b) The F-intercept is $(0, 32)$, so
$$^\circ\text{Fahrenheit} = 1.8(^\circ\text{Celsius}) + 32.$$

(c) If the temperature is 20° Celsius, then
$$^\circ\text{Fahrenheit} = 1.8(20) + 32 = 68^\circ\text{Fahrenheit}.$$

(d) If $^\circ$Fahrenheit $= ^\circ$Celsius, then
$$^\circ\text{Celsius} = 1.8^\circ\text{Celsius} + 32$$
$$-32 = 0.8^\circ\text{Celsius}$$
$$^\circ\text{Celsius} = -40^\circ = ^\circ\text{Fahrenheit}.$$

21. (a) Since the population center is moving west at 50 miles per 10 years, or 5 miles per year, if we start in 1990, when the center is in Steelville, its distance d west of Steelville t years after 1990 is given by

$$d = 5t.$$

(b) It moved over 700 miles over the 200 years from 1790 to 1990, so its average speed was greater than $\frac{700}{200} = 3.5$ miles/year, somewhat slower than its present rate.

(c) According to the function in (a), after 300 years the population center would be 1500 miles west of Steelville, in Baja, California, which seems rather unlikely.

Solutions for Section 1.3

1. The graph shows a concave up function.

5. Since $a = 7$ is greater than 1, this is an example of exponential growth.

9. We see that all ratios are the same:
$$\frac{h(5)}{h(4)} = \frac{h(6)}{h(5)} = \frac{h(7)}{h(6)} = \frac{1}{2}.$$

The function is decreasing as t increases, so this appears to be an example of exponential decay. To find a formula for $h(t)$, we let
$$h(t) = a \cdot b^t, \quad \text{with} \quad b = \frac{1}{2}.$$

Now, we solve for a:
$$2096 = a \cdot \left(\frac{1}{2}\right)^3$$
$$a = \frac{2096}{(1/2)^3} = 16768.$$

Thus, a possible formula for $h(t)$ is $h(t) = 16768(\frac{1}{2})^t$.

13.

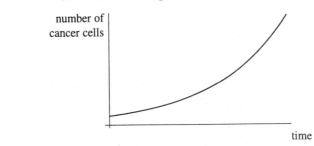

number of cancer cells

time

17. Since $(0.1)^t \to 0$ as $t \to \infty$, we see that $Z \to 3$ as $t \to \infty$. Therefore $Z = 3$ is a horizontal asymptote. Note $Z \to \infty$ as $t \to -\infty$.

21. We look for an equation of the form $y = y_0 a^x$ since the graph looks exponential. The points $(-1, 8)$ and $(1, 2)$ are on the graph, so
$$8 = y_0 a^{-1} \quad \text{and} \quad 2 = y_0 a^1$$

Therefore $\frac{8}{2} = \frac{y_0 a^{-1}}{y_0 a} = \frac{1}{a^2}$, giving $a = \frac{1}{2}$, and so $2 = y_0 a^1 = y_0 \cdot \frac{1}{2}$, so $y_0 = 4$.

Hence $y = 4\left(\frac{1}{2}\right)^x = 4(2^{-x})$.

25. We see that $\frac{1.09}{1.06} \approx 1.03$, and therefore $h(s) = c(1.03)^s$; c must be 1. Similarly $\frac{2.42}{2.20} = 1.1$, and so $f(s) = a(1.1)^s$; $a = 2$. Lastly, $\frac{3.65}{3.47} = 1.05$, so $g(s) = b(1.05)^s$; $b = 3$.

29. If the pressure at sea level is P_0, the pressure P at altitude h is given by
$$P = P_0 \left(1 - \frac{0.4}{100}\right)^{\frac{h}{100}},$$

since we want the pressure to be multiplied by a factor of $\left(1 - \frac{0.4}{100}\right) = 0.996$ for each 100 feet we go up to make it decrease by 0.4% over that interval. At Mexico City $h = 7340$, so the pressure is
$$P = P_0(0.996)^{\frac{7340}{100}} \approx 0.745 P_0.$$

So the pressure is reduced from P_0 to approximately $0.745 P_0$, a decrease of 25.5%.

Solutions for Section 1.4

1.   16

5.   1/3

9.   Exponential growth dominates power growth as $x \to \infty$, so $10 \cdot 2^x$ is larger.

13.   As $x \to \infty$, $y \to \infty$.
   As $x \to -\infty$, $y \to -\infty$.

17.

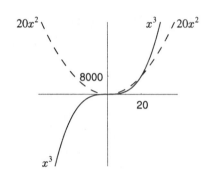

   $f(x) = x^3$ is larger as $x \to \infty$.

21.   Let $D(v)$ be the stopping distance required by an Alpha Romeo as a function of its velocity. The assumption that stopping distance is proportional to the square of velocity is equivalent to the equation

$$D(v) = kv^2$$

   where k is a constant of proportionality. To determine the value of k, we use the fact that $D(70) = 177$.

$$D(70) = k(70)^2 = 177.$$

Thus,

$$k = \frac{177}{70^2} \approx 0.0361.$$

It follows that

$$D(35) = \left(\frac{177}{70^2} \right)(35)^2 = \frac{177}{4} = 44.25 \text{ ft}$$

and

$$D(140) = \left(\frac{177}{70^2} \right)(140)^2 = 708 \text{ ft}.$$

   Thus at half the speed it requires one fourth the distance, whereas at twice the speed it requires four times the distance, as we would expect from the equation. (We could in fact have figured it out that way, without solving for k explicitly.)

25.   The graphs are shown in Figure 1.2.

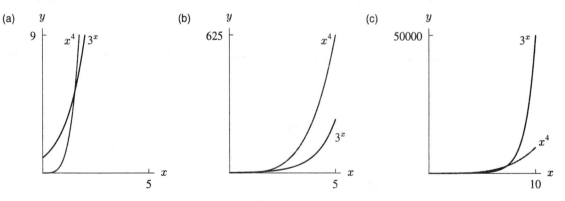

Figure 1.2

Solutions for Section 1.5

1. (a) $f(25)$ is q corresponding to $p = 25$, or, in other words, the number of items sold when the price is 25.
    (b) $f^{-1}(30)$ is p corresponding to $q = 30$, or the price at which 30 units will be sold.

5. Probably not invertible. Since your calculus class probably has less than 363 students, there will be at least two days in the year, say a and b, with $f(a) = f(b) = 0$. Hence we don't know what to choose for $f^{-1}(0)$.

9. The function is not invertible since there are horizontal lines which hit the function more than once.

13. (a) The function f tells us C in terms of q. To get its inverse, we want q in terms of C, which we find by solving for q:
$$C = 100 + 2q,$$
$$C - 100 = 2q,$$
$$q = \frac{C - 100}{2} = f^{-1}(C).$$
    (b) The inverse function tells us the number of articles that can be produced for a given cost.

17. (a) We find $f^{-1}(2)$ by finding the x value corresponding to $f(x) = 2$. Looking at the graph, we see that $f^{-1}(2) = -1$.
    (b) We construct the graph of $f^{-1}(x)$ by reflecting the graph of $f(x)$ over the line $y = x$. The graphs of $f^{-1}(x)$ and $f(x)$ are shown together in Figure 1.3.

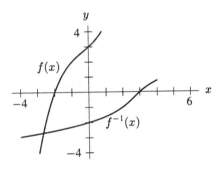

Figure 1.3

Solutions for Section 1.6

1.

x	1	2	3	4	5	6	7	8	9	10
$f(x)$	0	0.30	0.48	0.60	0.70	0.78	0.85	0.90	0.95	1.00
$g(x)$	1.00	1.41	1.73	2.00	2.24	2.45	2.65	2.83	3.00	3.16

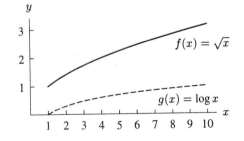

5. Isolating the exponential term

$$20 = 50(1.04)^x$$
$$\frac{20}{50} = (1.04)^x$$

Taking logs of both sides

$$\log \frac{2}{5} = \log(1.04)^x$$
$$\log \frac{2}{5} = x \log(1.04)$$
$$x = \frac{\log(2/5)}{\log(1.04)} \approx -23.4.$$

9. $t = \dfrac{\log a}{\log b}.$

13. Using the log rules, we have

$$3 \log A - \frac{2}{3} \log B + \frac{1}{3} \log A + \frac{5}{3} \log B = \frac{10}{3} \log A + \log B.$$

17. Using the log rules in the exponent, we have

$$100^{(\log A - \log B)} = 10^{2 \log \left(\frac{A}{B}\right)} = 10^{\log \left(\frac{A}{B}\right)^2} = \left(\frac{A}{B}\right)^2.$$

21. $\log x$ goes to infinity and $10^{-0.01x}$ goes to 0 as x goes to infinity.

25. Let $t =$ number of years since 1980. Then the number of vehicles, V, in millions, at time t is given by

$$V = 170(1.04)^t$$

and the number of people, P, in millions, at time t is given by

$$P = 227(1.01)^t.$$

There is an average of one vehicle per person when $\dfrac{V}{P} = 1$, or $V = P$. Thus, we must solve for t the equation:

$$170(1.04)^t = 227(1.01)^t,$$

which implies

$$\left(\frac{1.04}{1.01}\right)^t = \frac{(1.04)^t}{(1.01)^t} = \frac{227}{170}$$

Taking logs on both sides,

$$t \log \frac{1.04}{1.01} = \log \frac{227}{170}.$$

Therefore,

$$t = \frac{\log \left(\frac{227}{170}\right)}{\log \left(\frac{1.04}{1.01}\right)} \approx 9.9 \text{ years.}$$

So there was, according to this model, about one vehicle per person in 1990.

29. If $p(t) = (1.04)^t$, then, for p^{-1} the inverse of p, we should have

$$(1.04)^{p^{-1}(t)} = t,$$
$$p^{-1}(t) \log(1.04) = \log t,$$
$$p^{-1}(t) = \frac{\log t}{\log(1.04)} \approx 58.708 \log t.$$

Solutions for Section 1.7

1.

x	1	2	3	4	5	6	7	8	9	10
$f(x)$	0	0.30	0.48	0.60	0.70	0.78	0.85	0.90	0.95	1.00
$g(x)$	0	0.69	1.10	1.39	1.61	1.79	1.95	2.08	2.20	2.30

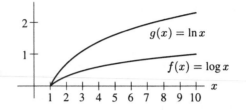

5. $$\ln(10^{x+3}) = \ln(5e^{7-x})$$
    $$(x+3)\ln 10 = \ln 5 + (7-x)\ln e$$
    $$2.303(x+3) = 1.609 + (7-x)$$
    $$3.303x = 1.609 + 7 - 2.303(3)$$
    $$x = 0.515$$

9. $\ln \dfrac{P}{P_0} = kt$, so $t = \dfrac{\ln \frac{P}{P_0}}{k}$.

13. Using the identity $\ln(e^x) = x$, we have 2AB.

17. $2\ln A + 2\ln B - \ln B - (-\ln A) = 3\ln A + \ln B$.

21. $P = 79(e^{-2.5})^t = 79(0.0821)^t$. Exponential decay because $-2.5 < 0$ or $0.0821 < 1$.

25. We want $1.7^t = e^{kt}$ so $1.7 = e^k$ and $k = \ln 1.7 = 0.5306$. Thus $P = 10e^{0.5306t}$.

29. Since f is increasing, f has an inverse. To find the inverse of $f(t) = 50e^{0.1t}$, we replace t with $f^{-1}(t)$, and, since $f(f^{-1}(t)) = t$, we have
    $$t = 50e^{0.1f^{-1}(t)}.$$
    We then solve for $f^{-1}(t)$:
    $$t = 50e^{0.1f^{-1}(t)}$$
    $$\frac{t}{50} = e^{0.1f^{-1}(t)}$$
    $$\ln\left(\frac{t}{50}\right) = 0.1f^{-1}(t)$$
    $$f^{-1}(t) = \frac{1}{0.1}\ln\left(\frac{t}{50}\right) = 10\ln\left(\frac{t}{50}\right).$$

33. (a) We have $P_0 = 1$ million, and $k = 0.02$, so $P = (1{,}000{,}000)(e^{0.02t})$.
    (b)

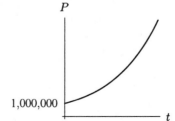

37.   Assuming the US population grows exponentially, we have

$$248.7 = 226.5e^{10k}$$
$$k = \frac{\ln(1.098)}{10} = 0.00935.$$

We want to find the time t in which

$$300 = 226.5e^{0.00935t}$$
$$t = \frac{\ln(1.324)}{0.00935} = 30 \text{ years.}$$

Thus, the population will go over 300 million around the year 2010.

41.   We can find the substance's decay constant k using the formula

$$0.5P_0 = P_0e^{-k(5)}$$
$$k = \frac{\ln(0.5)}{-5}.$$

(a)   Using this decay constant, we find that after $t = 10$ years

$$P = 20e^{\ln(0.5)10/5}$$
$$= 20e^{\ln(0.5^2)}$$
$$= 20(0.5^2)$$
$$= 5 \text{ kg.}$$

(b)   We wish to find t in which $P = 0.1$ kg, or

$$0.1 = 20e^{\ln(0.5)t/5}$$
$$t = 5\frac{\ln(0.1/20)}{\ln(0.5)} = 38.2 \text{ years.}$$

45. $e^{-k(5730)} = 0.5$ so $k = 1.21 \cdot 10^{-4}$. Thus, $e^{-1.21\cdot10^{-4}t} = 0.995$, so $t = \frac{\ln 0.995}{-1.21\cdot10^{-4}} = 41.43$ years, so the painting is a fake.

Solutions for Section 1.8

1.   Figures 1.4–1.7 show the appropriate graphs. Note that asymptotes are shown as dashed lines and x- or y-intercepts are shown as filled circles.

(a)

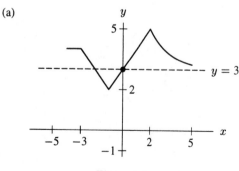

Figure 1.4

(b)

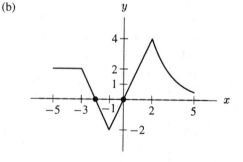

Figure 1.5

(c)

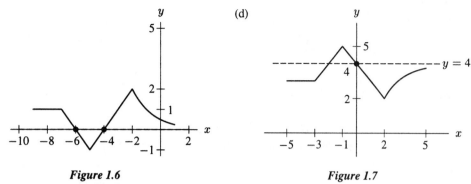

Figure 1.6

(d)

Figure 1.7

5. $f(h(x)) = f(e^{4x+7}) = 2e^{4x+7} + 1$

9. $g(h(x) - 3) = g(e^{4x+7} - 3) = \ln(e^{4x+7} - 3 + 3) = \ln(e^{4x+7}) = 4x + 7$

13.   The function is even since $f(-x) = f(x)$.

17.
$$f(-x) = (-x)^3 + (-x)^2 + (-x) = -x^3 + x^2 - x.$$
Since $f(-x) \neq f(x)$ and $f(-x) \neq -f(x)$, this function is neither even nor odd.

21. $f(x) = x^3, \quad g(x) = \ln x.$

25. $m(z + h) - m(z - h) = (z + h)^2 - (z - h)^2 = z^2 + 2hz + h^2 - (z^2 - 2hz + h^2) = 4hz.$

29.   (a)   We have $g(x) = f(x/2)$, and we can only evaluate $f(x/2)$ if $x/2$ is one of the x-values 0, 1, 2, 3, 4, 5, 6 in the table of values for f. So we can only evaluate $g(x)$ if x is one of 0, 2, 4, 6, 8, 10, 12.

     (b)   **TABLE 1.1**

x	0	2	4	6	8	10	12
$g(x)$	0	2	6	12	20	30	42

     (c)   The graph of g is obtained from that of f by a horizontal stretch by a factor of 2, so its graph will be twice as wide as f's.

33.   Computing $f(g(x))$ as in Problem 30, we get the following table. From it we graph $f(g(x))$.

x	$g(x)$	$f(g(x))$
-3	0.6	-0.5
-2.5	-1.1	-1.3
-2	-1.9	-1.2
-1.5	-1.9	-1.2
-1	-1.4	-1.3
-0.5	-0.5	-1
0	0.5	-0.6
0.5	1.4	-0.2
1	2	0.4
1.5	2.2	0.5
2	1.6	0
2.5	0.1	-0.7
3	-2.5	0.1

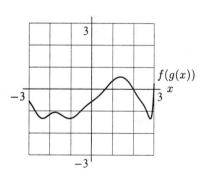

37.

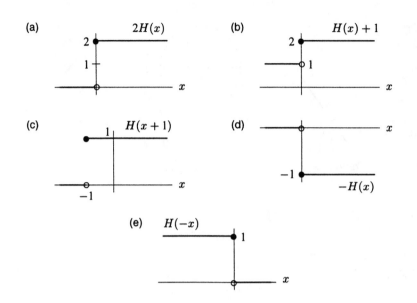

(a) $2H(x)$

(b) $H(x) + 1$

(c) $H(x + 1)$

(d) $-H(x)$

(e) $H(-x)$

41.

TABLE 1.2

x	$f(x)$	$g(x)$	$h(x)$
-3	0	0	0
-2	2	2	-2
-1	2	2	-2
0	0	0	0
1	2	-2	-2
2	2	-2	-2
3	0	0	0

Solutions for Section 1.9

1.

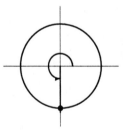

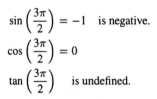

$$\sin\left(\frac{3\pi}{2}\right) = -1 \quad \text{is negative.}$$

$$\cos\left(\frac{3\pi}{2}\right) = 0$$

$$\tan\left(\frac{3\pi}{2}\right) \quad \text{is undefined.}$$

5.

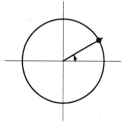

$$\sin\left(\frac{\pi}{6}\right) \text{ is positive.}$$

$$\cos\left(\frac{\pi}{6}\right) \text{ is positive.}$$

$$\tan\left(\frac{\pi}{6}\right) \text{ is positive.}$$

9. $-1 \text{ radian} \cdot \frac{180°}{\pi \text{ radians}} = -\left(\frac{180°}{\pi}\right) \approx -60°$

$$\sin(-1) \quad \text{is negative}$$

$$\cos(-1) \quad \text{is positive}$$

$$\tan(-1) \quad \text{is negative}$$

13.

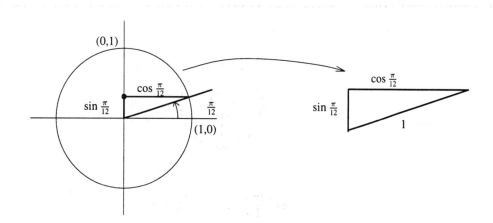

By the Pythagorean Theorem, $(\cos\frac{\pi}{12})^2 + (\sin\frac{\pi}{12})^2 = 1^2$; so $(\cos\frac{\pi}{12})^2 = 1 - (\sin\frac{\pi}{12})^2$ and $\cos\frac{\pi}{12} = \sqrt{1 - (\sin\frac{\pi}{12})^2} = \sqrt{1 - (0.259)^2} \approx 0.966$. We take the positive square root since by the picture we know that $\cos\frac{\pi}{12}$ is positive.

17. $\sin x^2$ is by convention $\sin(x^2)$, which means you square the x first and then take the sine.

$\sin^2 x = (\sin x)^2$ means find $\sin x$ and then square it.

$\sin(\sin x)$ means find $\sin x$ and then take the sine of that.

Expressing each as a composition: If $f(x) = \sin x$ and $g(x) = x^2$, then

$\sin x^2 = f(g(x))$

$\sin^2 x = g(f(x))$

$\sin(\sin x) = f(f(x))$.

21.   This graph is a sine curve with period 8π and amplitude 2, so it is given by $f(x) = 2\sin(\frac{x}{4})$.

25.   This graph has period 5, amplitude 1 and no vertical shift or horizontal shift from $\sin x$, so it is given by

$$f(x) = \sin\left(\frac{2\pi}{5}x\right).$$

29.
(a)   Beginning at time $t = 0$, the voltage will have oscillated through a complete cycle when $\cos(120\pi t) = \cos(2\pi)$, hence when $t = \frac{1}{60}$ second. The period is $\frac{1}{60}$ second.

(b) V_0 represents the amplitude of the oscillation.

(c)
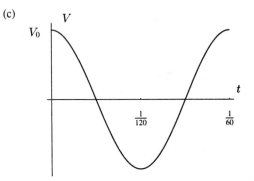

33.   (a)   A table of values for $f(x)$ is given below.

x	-1	-0.8	-0.6	-0.4	-0.2	0	0.2	0.4	0.6	0.8	1
arcsin x	-1.57	-0.93	-0.64	-0.41	-0.20	0	0.20	0.41	0.64	0.93	1.57

(b)   The domain is $-1 \le x \le 1$. The range is $-\frac{\pi}{2} \le x \le \frac{\pi}{2}$. A graph is shown in Figure 1.8.

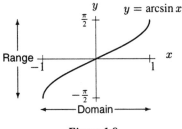

Figure 1.8

37.   The function R has period of π, so its graph is as shown in the figure below. The maximum value of the range is v_0^2/g and occurs when $\theta = \pi/4$.

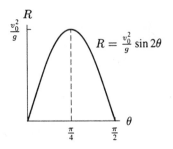

Solutions for Section 1.10

1. (a) Since $f(x)$ is an odd polynomial with a positive leading coefficient, it follows that $f(x) \to \infty$ as $x \to \infty$ and $f(x) \to -\infty$ as $x \to -\infty$.
   (b) Since $f(x)$ is an even polynomial with negative leading coefficient, it follows that $f(x) \to -\infty$ as $x \to \pm\infty$.
   (c) As $x \to \pm\infty$, $x^4 \to \infty$, so $x^{-4} = 1/x^4 \to 0$.
   (d) As $x \to \pm\infty$, the lower-degree terms of $f(x)$ become insignificant, and $f(x)$ becomes approximated by the highest degree terms in its numerator and denominator. So as $x \to \pm\infty$, $f(x) \to 6$.

5.

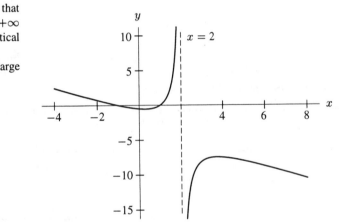

9. To find vertical asymptote(s), look at the behavior of y as x approaches a value for which the denominator is 0.

$$x - 2 = 0 \quad \text{when} \quad x = 2.$$

If we plug in values for x near 2, we will see that $y \to -\infty$ as $x \to 2$ from the right, but $y \to +\infty$ as $x \to 2$ from the left. Clearly, $x = 2$ is a vertical asymptote.

$y(x)$ has no horizontal asymptote because, for large x,

$$y = \frac{1 - x^2}{x - 2} \approx \frac{-x^2}{x} = -x,$$

and so

$$y \to -\infty \quad \text{as} \quad x \to +\infty,$$
$$y \to +\infty \quad \text{as} \quad x \to -\infty.$$

(Also, if $|x|$ is big enough, $\dfrac{1 - x^2}{x - 2} \approx -x$, and the straight line $y = -x$ is an asymptote for the graph.)

13. (a) The object starts at $t = 0$, when $s = v_0(0) - g(0)^2/2 = 0$. Thus it starts on the ground, with zero height.
    (b) The object hits the ground when $s = 0$. This is satisfied at $t = 0$, before it has left the ground, and at some later time t that we must solve for.
    $$0 = v_0 t - gt^2/2 = t\left(v_0 - gt/2\right)$$
    Thus $s = 0$ when $t = 0$ and when $v_0 - gt/2 = 0$, i.e., when $t = 2v_0/g$. The starting time is $t = 0$, so it must hit the ground at time $t = 2v_0/g$.
    (c) The object reaches its maximum height halfway between when it is released and when it hits the ground, or at
    $$t = (2v_0/g)/2 = v_0/g.$$
    (d) Since we know the time at which the object reaches its maximum height, to find the height it actually reaches we just use the given formula, which tells us s at any given t. Plugging in $t = \frac{v_0}{g}$,
    $$s = v_0\left(\frac{v_0}{g}\right) - \frac{1}{2}g\left(\frac{v_0^2}{g^2}\right) = \frac{v_0^2}{g} - \frac{v_0^2}{2g}$$
    $$= \frac{2v_0^2 - v_0^2}{2g} = \frac{v_0^2}{2g}.$$

17.   (a)   Because our cubic has a root at 2 and a double root at -2, it has the form

$$y = k(x + 2)(x + 2)(x - 2).$$

Since $y = 4$ when $x = 0$,

$$4 = k(2)(2)(-2) = -8k,$$
$$k = -\frac{1}{2}.$$

Thus our equation is

$$y = -\frac{1}{2}(x + 2)^2(x - 2).$$

21.   (a) $f(x) = k(x + 2)(x - 2)^2(x - 5) = k(x^4 - 7x^3 + 6x^2 + 28x - 40)$, where $k < 0$. ($k \approx -\frac{1}{15}$ if the scales are equal; otherwise one can't tell how large k is.)

(b)   This function is increasing for $x < -1$ and for $2 < x < 4$, decreasing for $-1 < x < 2$ and for $4 < x$.

25.   (a) $a(v) = \frac{1}{m}(\text{ENGINE} - \text{WIND}) = \frac{1}{m}(F_E - kv^2)$, where k is a positive constant.

(b)   A possible graph is shown below.

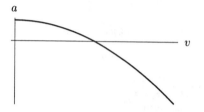

29.   Since the parabola opens upward, we must have $a > 0$. To determine a relationship between x and y at the point of intersection P, we eliminate a from the parabola and circle equations. Since $y = x^2/a$, we have $a = x^2/y$. Putting this into the circle equation gives $x^2 + y^2 = 2x^4/y^2$. Rewrite this as

$$x^2 y^2 + y^4 = 2x^4$$
$$y^4 + x^2 y^2 - 2x^4 = 0$$
$$(y^2 + 2x^2)(y^2 - x^2) = 0.$$

This means $x^2 = y^2$ (since y^2 cannot equal $-2x^2$). Thus $x = y$ since P is in the first quadrant. So P moves out along the line $y = x$ through the origin.

Solutions for Section 1.11

1.   No, because $x - 2 = 0$ at $x = 2$.

5.   No, because $\sin 0 = 0$.

9.   The graph of g suggests that g is not continuous on any interval containing $\theta = 0$, since $g(0) = 1/2$.

Solutions for Chapter 1 Review

1.   (a)   The domain of f is the set of values of x for which the function is defined. Since the function is defined by the graph and the graph goes from $x = 0$ to $x = 7$, the domain of f is $[0, 7]$.

(b)   The range of f is the set of values of y attainable over the domain. Looking at the graph, we can see that y gets as high as 5 and as low as -2, so the range is $[-2, 5]$.

(c)   Only at $x = 5$ does $f(x) = 0$. So 5 is the only root of $f(x)$.

(d)   Looking at the graph, we can see that $f(x)$ is decreasing on $(1, 7)$.

(e)   The graph indicates that $f(x)$ is concave up at $x = 6$.

(f)   The value $f(4)$ is the y-value that corresponds to $x = 4$. From the graph, we can see that $f(4)$ is approximately 1.

(g)   This function is not invertible, since it fails the horizontal-line test. A horizontal line at $y = 3$ would cut the graph of $f(x)$ in two places, instead of the required one.

5. (a) Advertising is generally cheaper in bulk; spending more money will give better and better marginal results initially, (Spending $5,000 could give you a big newspaper ad reaching 200,000 people; spending $100,000 could give you a series of TV spots reaching 50,000,000 people.) A graph is shown below, left.

(b) The temperature of a hot object decreases at a rate proportional to the difference between its temperature and the temperature of the air around it. Thus, the temperature of a very hot object decreases more quickly than a cooler object. The graph is decreasing and concave up. (We are assuming that the coffee is all at the same temperature.)

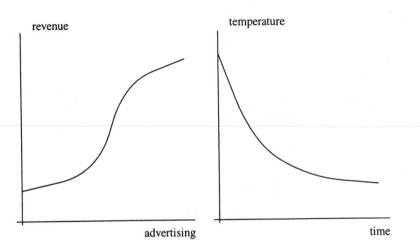

9. From the figure it appears that the "seat" of the graph $y = x^3$ has been moved to the left and up by 1, to the point $(-1, 1)$. Since translation to the right by h is achieved by replacing x with $(x - h)$ and translation up by k is achieved by replacing y with $(y - k)$, the equation of our translated graph appears to be

$$y - 1 = (x - (-1))^3$$

or

$$y = (x + 1)^3 + 1.$$

Since the picture *suggests* that the graph of $y = x^3$ has been moved over and up by 1 (but does not show this explicitly), we will confirm that our equation is correct by checking the x and y intercepts (which are shown explicitly in the picture). The desired y-intercept is 2, and substituting $x = 0$ into the equation gives

$$y = (0 + 1)^3 + 1 = 2.$$

In addition, the desired x-intercept is -2, and substituting $y = 0$ into the equation gives

$$0 = (x + 1)^3 + 1$$

so

$$(x + 1)^3 = -1$$
$$x + 1 = \sqrt[3]{-1} = -1$$
$$x = -2.$$

Thus our equation does have the graph shown.

We could also have solved this problem, with less guessing and more algebra, by finding a translation of x^3 which has the given x- and y-intercepts. Let h be the horizontal, and k the vertical translation; then the formula for the translated function is

$$f(x) = (x - h)^3 + k.$$

The y-intercept of f is $f(0) = 2$, so

$$f(0) = (0 - h)^3 + k = -h^3 + k = 2.$$

Therefore $k = 2 + h^3$ and we have $f(x) = (x - h)^3 + (2 + h^3)$. The x-intercept, -2, is the x-value for which $f(x) = 0$, so

$$
\begin{aligned}
0 = f(-2) &= (-2 - h)^3 + (2 + h^3) \\
&= -6h^2 - 12h - 6 \\
&= -6(h + 1)^2
\end{aligned}
$$

Therefore $h = -1$ and $k = 2 + h^3 = 1$.

13. We will let

$$
\begin{aligned}
T &= \text{ amount of fuel for take-off} \\
L &= \text{ amount of fuel for landing} \\
P &= \text{ amount of fuel per mile in the air} \\
m &= \text{ the length of the trip in miles}
\end{aligned}
$$

Then Q, the total amount of fuel needed, is given by

$$
Q(m) = T + L + Pm
$$

17. (a) $f(n) + g(n) = (3n^2 - 2) + (n + 1) = 3n^2 + n - 1$.
    (b) $f(n)g(n) = (3n^2 - 2)(n + 1) = 3n^3 + 3n^2 - 2n - 2$.
    (c)   the domain of $\frac{f(n)}{g(n)}$ is defined everywhere where $g(n) \neq 0$, i.e. for all $n \neq -1$.
    (d) $f(g(n)) = 3(n + 1)^2 - 2 = 3n^2 + 6n + 1$.
    (e) $g(f(n)) = (3n^2 - 2) + 1 = 3n^2 - 1$.

21. The graph appears to have a vertical asymptote at $t = 0$, so $f(t)$ is not continuous on $[-1, 1]$.

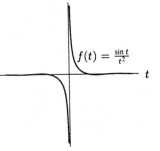

25. (a)   Since $Q = 25$ at $t = 0$, we have $Q_0 = 25$ (since $e^0 = 1$). We then plug the value at $t = 1$ into the equation

$$
Q = 25e^{rt}
$$

to find r. Doing so, we get

$$
\begin{aligned}
43 &= 25e^{r(1)} \\
\frac{43}{25} &= 1.72 = e^r \\
\ln 1.72 &= r \\
0.5423 &\approx r.
\end{aligned}
$$

And so the equation is

$$
Q = 25e^{0.5423t}.
$$

(b)   At the time t when the population has doubled,

$$
\begin{aligned}
2 &= e^{0.5423t} \\
\ln 2 &= 0.5423t \\
t &= \frac{\ln 2}{0.5423} \approx 1.3 \text{ months.}
\end{aligned}
$$

(c)   At the time t when the population is 1000 rabbits,

$$1000 = 25e^{0.5423t}$$
$$40 = e^{0.5423t}$$
$$\ln 40 = 0.5423t$$
$$t = \frac{\ln 40}{0.5423} \approx 6.8 \text{ months.}$$

29.   Using the exponential decay equation $P = P_0 e^{-kt}$, we can solve for the substance's decay constant k:

$$(P_0 - 0.3P_0) = P_0 e^{-20k}$$
$$k = \frac{\ln(0.7)}{-20}.$$

Knowing this decay constant, we can solve for the half-life t using the formula

$$0.5P_0 = P_0 e^{\ln(0.7)t/20}$$
$$t = \frac{20\ln(0.5)}{\ln(0.7)} \approx 38.87 \text{ hours.}$$

33.   The period T_E of the earth is (by definition!) one year or about 365.24 days. Since the semimajor axis of the earth is 150 million km, we can use Kepler's Law to derive the constant of proportionality, k.

$$T_E = k(S_E)^{\frac{3}{2}}$$

where S_E is the earth's semimajor axis, or 150 million km.

$$365.24 = k(150)^{\frac{3}{2}}$$

$$k = \frac{365.24}{(150)^{\frac{3}{2}}} \approx 0.198.$$

Now that we know the constant of proportionality, we can use it to derive the periods of Mercury and Pluto. For Mercury,

$$T_M = (0.198)(58)^{\frac{3}{2}} \approx 87.818 \text{ days.}$$

For Pluto,

$$T_P = (0.198)(6000)^{\frac{3}{2}} \approx 92,400 \text{ days,}$$

or (converting Pluto's period to years),

$$\frac{(0.198)(6000)^{\frac{3}{2}}}{365.24} \approx 253 \text{ years.}$$

37.   There are many solutions for a graph like this one. The simplest is $y = 1 - e^{-x}$, which gives the graph of $y = e^x$, flipped over the x-axis and moved up by 1. The resulting graph passes through the origin and approaches $y = 1$ as an upper bound, the two features of the given graph.

41. $y = 5\sin\left(\frac{\pi t}{20}\right)$

45.   (a)   (i)   The inverse function is shown in Figure 1.9.

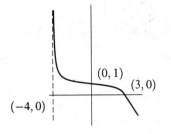

$(0, 1)$

$(3, 0)$

$(-4, 0)$

Figure 1.9

(ii)   The reciprocal function is shown in Figure 1.10.

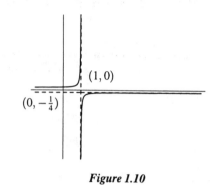

Figure 1.10

(b)   The horizontal asymptote at $y = -4$ in the original function becomes a vertical asymptote at $x = -4$ in the inverse function.

49.   (a)   The rate R is the difference of the rate at which the glucose is being injected, which is given to be constant, and the rate at which the glucose is being broken down, which is given to be proportional to the amount of glucose present. Thus we have the formula

$$R = k_1 - k_2 G$$

where k_1 is the rate that the glucose is being injected, k_2 is the constant relating the rate that it is broken down to the amount present, and G is the amount present.

(b)

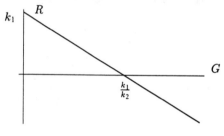

Solutions to Problems on the Binomial Theorem

1.   From the formula, C_k^n looks as though it could be a fraction; however, the entries we see in the first 6 rows of Pascal's triangle are all integers. Since all further entries are obtained by adding the entries above, they must all be positive integers.

Alternatively, C_k^n represents the number of ways the term $x^{n-k}y^k$ arises when $(x + y)^n$ is multiplied out. This gives another way of seeing that C_k^n must be a positive integer.

Solutions to Problems on the Completeness of Real Numbers

1.   (a)   (i)   A lower bound is a number which is less than or equal to all the numbers in the set.

(ii)   The greatest lower bound of a set is the one which is greater than or equal to all the others.

(b)   Any nonempty set of real numbers which has a lower bound has a greatest lower bound.

CHAPTER TWO

Solutions for Section 2.1

1.

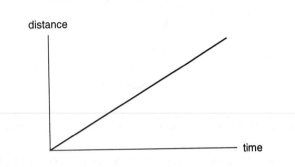

5.

TABLE 2.1

slope	point
-3	F
-1	C
0	E
$\frac{1}{2}$	A
1	B
2	D

9.

$$\frac{(3+0.1)^3 - 27}{0.1} = 27.91$$

$$\frac{(3+0.01)^3 - 27}{0.01} = 27.09$$

$$\frac{(3+0.001)^3 - 27}{0.001} = 27.009$$

These calculations suggest that $\lim_{h \to 0} \dfrac{(3+h)^3 - 27}{h} = 27$

13.

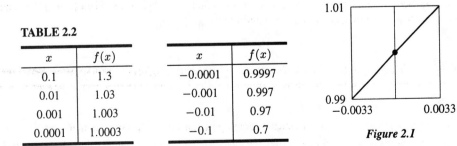

TABLE 2.2

x	$f(x)$	x	$f(x)$
0.1	1.3	-0.0001	0.9997
0.01	1.03	-0.001	0.997
0.001	1.003	-0.01	0.97
0.0001	1.0003	-0.1	0.7

Figure 2.1

From Table 2.2, it appears the limit is 1. This is confirmed by Figure 2.1. An appropriate window is $-0.0033 < x < 0.0033$, $0.99 < y < 1.01$.

17.

TABLE 2.3

x	$f(x)$
0.1	1.9867
0.01	1.9999
0.001	2.0000
0.0001	2.0000

x	$f(x)$
−0.0001	2.0000
−0.001	2.0000
−0.01	1.9999
−0.1	1.9867

Figure 2.2

From Table 2.3, it appears the limit is 2. This is confirmed by Figure 2.2. An appropriate window is $-0.0865 < x < 0.0865$, $1.99 < y < 2.01$.

21.

TABLE 2.4

x	$f(x)$
0.1	0.0666
0.01	0.0067
0.001	0.0007
0.0001	0

x	$f(x)$
−0.0001	−0.0001
−0.001	−0.0007
−0.01	−0.0067
−0.1	−0.0666

Figure 2.3

From Table 2.4, it appears the limit is 0. This is confirmed by Figure 2.3. An appropriate window is $-0.015 < x < 0.015$, $-0.01 < y < 0.01$.

Solutions for Section 2.2

1.

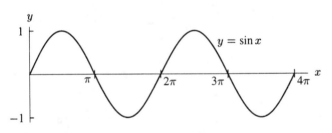

Since $\sin x$ is decreasing for values near $x = 3\pi$, its derivative at $x = 3\pi$ is negative.

5.

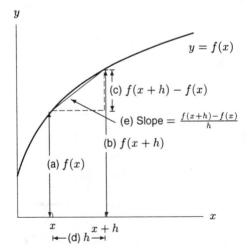

9. Figure 2.4 shows the quantities in which we are interested.

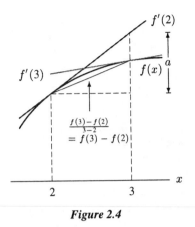

Figure 2.4

The quantities $f'(2)$, $f'(3)$ and $f(3) - f(2)$ have the following interpretations:

- $f'(2)$ = slope of the tangent line at $x = 2$
- $f'(3)$ = slope of the tangent line at $x = 3$
- $f(3) - f(2) = \frac{f(3)-f(2)}{3-2}$ = slope of the secant line from $f(2)$ to $f(3)$.

From Figure 2.4, it is clear that $0 < f(3) - f(2) < f'(2)$. By extending the secant line past the point $(3, f(3))$, we can see that it lies above the tangent line at $x = 3$. Thus $0 < f'(3) < f(3) - f(2) < f'(2)$. From the figure, the height a appears less than 1, so $f'(2) = \frac{a}{3-2} = \frac{a}{1} < 1$.

Thus

$$0 < f'(3) < f(3) - f(2) < f'(2) < 1.$$

13. Using the definition of the derivative

$$
\begin{aligned}
g'(-1) &= \lim_{h \to 0} \frac{g(-1+h) - g(-1)}{h} \\
&= \lim_{h \to 0} \frac{(3(-1+h)^2 + 5(-1+h)) - (3(-1)^2 + 5(-1))}{h} \\
&= \lim_{h \to 0} \frac{(3(1 - 2h + h^2) - 5 + 5h) - (-2)}{h} \\
&= \lim_{h \to 0} \frac{3 - 6h + 3h^2 - 3 + 5h}{h} \\
&= \lim_{h \to 0} \frac{(-h + 3h^2)}{h} = \lim_{h \to 0} (-1 + 3h) = -1.
\end{aligned}
$$

17.

$$
\begin{aligned}
g'(2) &= \lim_{h \to 0} \frac{g(2+h) - g(2)}{h} = \lim_{h \to 0} \frac{\frac{1}{2+h} - \frac{1}{2}}{h} \\
&= \lim_{h \to 0} \frac{2 - (2+h)}{h(2+h)2} = \lim_{h \to 0} \frac{-h}{h(2+h)2} \\
&= \lim_{h \to 0} \frac{-1}{(2+h)2} = -\frac{1}{4}
\end{aligned}
$$

21. We know that the slope of the tangent line to $f(x) = x$ when $x = 20$ is 1. When $x = 20$, $f(x) = 20$ so $(20, 20)$ is on the tangent line. Thus the equation of the tangent line is $y = 1(x - 20) + 20 = x$.

25.   (a)

$$f'(0) = \lim_{h \to 0} \frac{\overbrace{\sin h}^{h \text{ in degrees}} - \overbrace{\sin 0}^{0}}{h} = \frac{\sin h}{h}.$$

To four decimal places,

$$\frac{\sin 0.2}{0.2} \approx \frac{\sin 0.1}{0.1} \approx \frac{\sin 0.01}{0.01} \approx \frac{\sin 0.001}{0.001} \approx 0.01745$$

so $f'(0) \approx 0.01745$.

(b)   Consider the ratio $\frac{\sin h}{h}$. As we approach 0, the numerator, $\sin h$, will be much smaller in magnitude if h is in degrees than it would be if h were in radians. For example, if $h = 1°$ radian, $\sin h = 0.8415$, but if $h = 1$ degree, $\sin h = 0.01745$. Thus, since the numerator is smaller for h measured in degrees while the denominator is the same, we expect the ratio $\frac{\sin h}{h}$ to be smaller.

29.   (a)   We construct the difference quotient using erf(0) and each of the other given values:

$$\text{erf}'(0) \approx \frac{\text{erf}(1) - \text{erf}(0)}{1 - 0} = 0.84270079$$

$$\text{erf}'(0) \approx \frac{\text{erf}(0.1) - \text{erf}(0)}{0.1 - 0} = 1.1246292$$

$$\text{erf}'(0) \approx \frac{\text{erf}(0.01) - \text{erf}(0)}{0.01 - 0} = 1.128342.$$

Based on these estimates, the best estimate is $\text{erf}'(0) \approx 1.12$; the subsequent digits have not yet stabilized.

(b)   Using erf(0.001), we have

$$\text{erf}'(0) \approx \frac{\text{erf}(0.001) - \text{erf}(0)}{0.001 - 0} = 1.12838$$

and so the best estimate is now 1.1283.

33.   As h gets smaller, round-off error becomes important. When $h = 10^{-12}$, the quantity $2^h - 1$ is so close to 0 that the calculator rounds off the difference to 0, making the difference quotient 0. The same thing will happen when $h = 10^{-20}$.

Solutions for Section 2.3

1.   The graph is that of the line $y = -2x + 2$. Its derivative is -2.

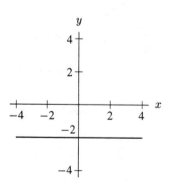

5.

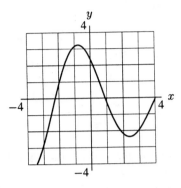

9.

13. Using the definition of the derivative,

$$g'(x) = \lim_{h \to 0} \frac{g(x+h) - g(x)}{h} = \lim_{h \to 0} \frac{2(x+h)^2 - 3 - (2x^2 - 3)}{h}$$

$$= \lim_{h \to 0} \frac{2(x^2 + 2xh + h^2) - 3 - 2x^2 + 3}{h} = \lim_{h \to 0} \frac{4xh + 2h^2}{h}$$

$$= \lim_{h \to 0} (4x + 2h) = 4x.$$

17. From the given information we know that f is increasing for values of x less than -2, is decreasing between $x = -2$ and $x = 2$, and is constant for $x > 2$. Figure 2.5 shows a possible graph—yours may be different.

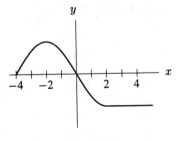

Figure 2.5

21.

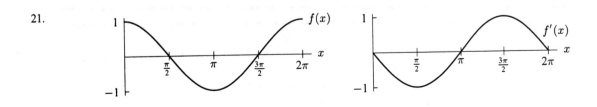

25.

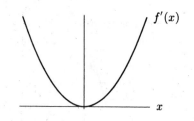

29.   (a) x_3 (b) x_4 (c) x_5 (d) x_3

33.   If $f(x)$ is even, its graph is symmetric about the y-axis. So the tangent line to f at $x = x_0$ is the same as that at $x = -x_0$ reflected about the y-axis.

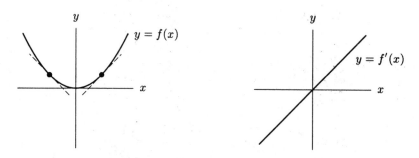

So the slopes of these two tangent lines are opposite in sign, so $f'(x_0) = -f'(-x_0)$, and f' is odd.

Solutions for Section 2.4

1.   The units of $f'(x)$ are feet/mile. The derivative, $f'(x)$, represents the rate of change of elevation with distance from the source, so if the river is flowing downhill everywhere, the elevation is always decreasing and $f'(x)$ is always negative. (In fact, there may be some stretches where the elevation is more or less constant, so $f'(x) = 0$.)

5.   Since B is measured in dollars and t is measured in years, $\frac{dB}{dt}$ is measured in dollars per year. We can interpret dB as the extra money added to your balance in dt years. Therefore $\frac{dB}{dt}$ represents how fast your balance is growing, in units of dollars/year.

9.   (a)   The pressure in dynes/cm^2 at a depth of 100 meters.
       (b)   The depth of water in meters giving a pressure of $1.2 \cdot 10^6$ dynes/cm^2.
       (c)   The pressure at a depth of h meters plus a pressure of 20 dynes/cm^2.
       (d)   The pressure at a depth of 20 meters below the diver.
       (e)   The rate of increase of pressure with respect to depth, at 100 meters, in units of dynes/cm^2 per meter. Approximately, $p'(100)$ represents the increase in pressure in going from 100 meters to 101 meters.
       (f)   The depth, in meters, at which the rate of change of pressure with respect to depth is 20 dynes/cm^2 per meter.

13.   Units of $g'(55)$ are mpg/mph. $g'(55) = -0.54$ means that at 55 miles per hour the fuel efficiency (in miles per gallon, or mpg) of a car decreases as the velocity increases at a rate of approximately one half mpg for an increase of one mph.

Solutions for Section 2.5

1. $f'(x) > 0$
      $f''(x) > 0$

5. $f'(x) > 0$
      $f''(x) < 0$

9. (a)

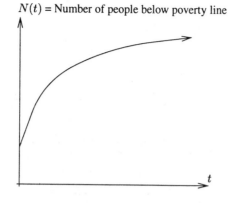

$N(t)$ = Number of people below poverty line

(b) $\frac{dN}{dt}$ is positive, since people are still slipping below the poverty line. $\frac{d^2N}{dt^2}$ is negative, since the rate at which people are slipping below the poverty line, $\frac{dN}{dt}$, is decreasing.

13. (a) The EPA will say that the rate of discharge is still rising. The industry will say that the rate of discharge is increasing less quickly, and may soon level off or even start to fall.

    (b) The EPA will say that the rate at which pollutants are being discharged is levelling off, but not to zero — so pollutants will continue to be dumped in the lake. The industry will say that the rate of discharge has decreased significantly.

17. (a) B and E   (b) A and D

Solutions for Chapter 2 Review

1.

distance

time

5.

x	$\ln x$	x	$\ln x$	x	$\ln x$	x	$\ln x$
0.998	−0.0020	1.998	0.6921	4.998	1.6090	9.998	2.3024
0.999	−0.0010	1.999	0.6926	4.999	1.6092	9.999	2.3025
1.000	0.0000	2.000	0.6931	5.000	1.6094	10.000	2.3026
1.001	0.0010	2.001	0.6936	5.001	1.6096	10.001	2.3027
1.002	0.0020	2.002	0.6941	5.002	1.6098	10.002	2.3028

At $x = 1$, the values of $\ln x$ are increasing by 0.001 for each increase in x of 0.001, so the derivative appears to be 1. At $x = 2$, the increase is 0.0005 for each increase of 0.001, so the derivative appears to be 0.5. At $x = 5$, $\ln x$ increases by 0.0002 for each increase of 0.001 in x, so the derivative appears to be 0.2. And at $x = 10$, the increase is 0.0001 over intervals of 0.001, so the derivative appears to be 0.1. These values suggest an inverse relationship between x and $f'(x)$, namely $f'(x) = \frac{1}{x}$.

9.

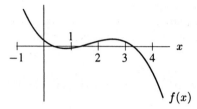

13. Using the definition of the derivative

$$f'(x) = \lim_{h \to 0} \frac{f(x+h) - f(x)}{h}$$

$$= \lim_{h \to 0} \frac{5(x+h)^2 + x + h - (5x^2 + x)}{h}$$

$$= \lim_{h \to 0} \frac{5(x^2 + 2xh + h^2) + x + h - 5x^2 - x}{h}$$

$$= \lim_{h \to 0} \frac{10xh + 5h^2 + h}{h}$$

$$= \lim_{h \to 0} (10x + 5h + 1) = 10x + 1$$

17. (a) The yam is cooling off so T is decreasing and $f'(t)$ is negative.

   (b) Since $f(t)$ is measured in degrees Fahrenheit and t is measured in minutes, df/dt must be measured in units of $F°/\text{min}$.

21. (a) See (b).

   (b)

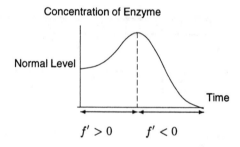

   (c) f' is the rate at which the concentration is increasing or decreasing. f' is positive at the start of the disease and negative toward the end. In practice, of course, one cannot measure f' directly. Checking the value of C in blood samples taken on consecutive days would tell us

$$\frac{f(t+1) - f(t)}{(t+1) - t},$$

   which is our estimate of $f'(t)$.

25. (a)

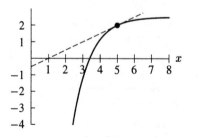

   (b) Exactly one. There can't be more than one zero because f is increasing everywhere. There does have to be one zero because f stays below its tangent line (dotted line in above graph), and therefore f must cross the x-axis.

(c) The equation of the (dotted) tangent line is $y = \frac{1}{2}x - \frac{1}{2}$, and so it crosses the x-axis at $x = 1$. Therefore the zero of f must be between $x = 1$ and $x = 5$.

(d) $\lim\limits_{x \to -\infty} f(x) = -\infty$, because f is increasing and concave down. Thus, as $x \to -\infty$, $f(x)$ decreases, at a faster and faster rate.

(e) Yes.

(f) No. The slope is decreasing since f is concave down, so $f'(1) > f'(5)$, i.e. $f'(1) > \frac{1}{2}$.

Solutions to Problems on Limits and Continuity

1. The graph in Figure 2.6 suggests that

$$\text{if } -0.05 < \theta < 0.05, \quad \text{then} \quad 0.999 < (\sin \theta)/\theta < 1.001.$$

Thus, if θ is within 0.05 of 0, we see that $(\sin \theta)/\theta$ is within 0.001 of 1.

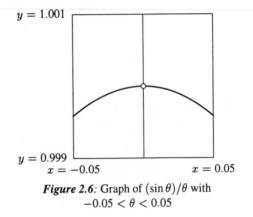

Figure 2.6: Graph of $(\sin \theta)/\theta$ with $-0.05 < \theta < 0.05$

5. For $-90° \leq \theta \leq 90°$, $0 \leq y \leq 0.02$, the graph of $y = \dfrac{\sin \theta}{\theta}$ is shown in Figure 2.7. Therefore, by tracing along the curve, we see that in degrees, $\lim\limits_{\theta \to 0} \dfrac{\sin \theta}{\theta} = 0.01745 \ldots$.

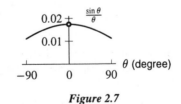

Figure 2.7

9. (a) Since $\sin(n\pi) = 0$ for $n = 1, 2, 3, \ldots$ the sequence of x-values

$$\frac{1}{\pi}, \frac{1}{2\pi}, \frac{1}{3\pi}, \ldots$$

works. These x-values $\to 0$ and are zeroes of $f(x)$.

(b) Since $\sin(n\pi/2) = 1$ for $n = 1, 5, 9 \ldots$ the sequence of x-values

$$\frac{2}{\pi}, \frac{2}{5\pi}, \frac{2}{9\pi}, \ldots$$

works.

(c) Since $\sin(n\pi)/2 = -1$ for $n = 3, 7, 11, \ldots$ the sequence of x-values

$$\frac{2}{3\pi}, \frac{2}{7\pi}, \frac{2}{11\pi} \ldots$$

works.

(d) Any two of these sequences of x-values show that if the limit were to exist, then it would have to have two (different) values: 0 and 1, or 0 and -1, or 1 and -1. Hence, the limit can not exist.

13. (a) If $b = 0$, then the property says $\lim_{x \to c} 0 = 0$, which is easy to see is true.

(b) If $|f(x) - L| < \frac{\epsilon}{|b|}$, then multiplying by $|b|$ gives

$$|b||f(x) - L| < \epsilon.$$

Since

$$|b||f(x) - L| = |b(f(x) - L)| = |bf(x) - bL|,$$

we have

$$|bf(x) - bL| < \epsilon.$$

(c) Suppose that $\lim_{x \to c} f(x) = L$. We want to show that $\lim_{x \to c} bf(x) = bL$. If we are to have

$$|bf(x) - bL| < \epsilon,$$

then we will need

$$|f(x) - L| < \frac{\epsilon}{|b|}.$$

We choose δ small enough that

$$|x - c| < \delta \quad \text{implies} \quad |f(x) - L| < \frac{\epsilon}{|b|}.$$

By part (b), this ensures that

$$|bf(x) - bL| < \epsilon,$$

as we wanted.

17. We can use the δ guaranteed by the continuity of f to be the "ϵ" for the continuity of g. That is, we can choose $\delta_1 > 0$ so that when $|x - c| < \delta_1$ we have $|g(x) - g(c)| < \delta$. But if we let $g(x) = y$ and remember $g(c) = d$, then this says that $|x - c| < \delta_1$ implies $|y - d| < \delta$, which in turn implies $|f(y) - f(d)| < \epsilon$. So, given any $\epsilon > 0$, we can find a $\delta_1 > 0$ such that $|x - c| < \delta_1$ implies $|f(y) - f(d)| = \left| f\left(g(x)\right) - f\left(g(c)\right)\right| < \epsilon$. This proves that $\lim_{x \to c} f\left(g(x)\right) = f(g(c))$, which is what it means for $f\left(g(x)\right)$ to be continuous at $x = c$

21. For any $\epsilon > 0$, we want to find the δ such that

$$|g(x) - 2| = \left|-x^3 + 2 - 2\right| = \left|x^3\right| < \epsilon.$$

Choose $\delta = \epsilon^{1/3}$. Then if $|x| < \delta = \epsilon^{1/3}$, it follows that $|g(x) - 2| = \left|x^3\right| < \epsilon$.

25. Since $\sin x$ is continuous everywhere, and $1/x$ is continuous except at $x = 0$, we know that $\sin(1/x)$ is continuous at each $x \neq 0$, because the composition of continuous functions is continuous. Furthermore, the product of two continuous functions is continuous, so $f(x)$ is continuous except perhaps at $x = 0$. Thus the only place we need to check continuity is $x = 0$. Since the values of the sine function go between 1 and -1, we have $|\sin(1/x)| \leq 1$ for $x \neq 0$, so

$$\left| x \sin\left(\frac{1}{x}\right)\right| \leq |x|.$$

It follows that we can make $x \sin(1/x)$ as small as we like by making x small enough. Hence $\lim_{x \to 0} (x \sin(1/x)) = 0$, so

$$\lim_{x \to 0} f(x) = \lim_{x \to 0} \left(x \sin\left(\frac{1}{x}\right)\right) = 0 = f(0).$$

So f is continuous at $x = 0$.

On the other hand, $\sin(1/x)$ has infinitely many oscillations between ϵ and 0, since it crosses the x-axis infinitely many times, at the points $x = 1/(n\pi)$, where n is an integer. So $x \sin(1/x)$ also has infinitely many oscillations, although they get smaller and smaller as $x \to 0$. Thus, f is not always increasing or always decreasing on any interval of the form $[0, \epsilon]$.

Solutions to Problems on Differentiability and Linear Approximation ▬▬▬

1. (a) (i) Function f is not continuous at $x = 1$.

(ii) Function f appears not differentiable at $x = 1, 2, 3$.

(b) (i) Function g appears continuous at all x-values shown.

(ii) Function g appears not differentiable at $x = 2, 4$.

5. We can see from Figure 2.8 that the graph of f oscillates infinitely often between the curves $y = x^2$ and $y = -x^2$ near the origin. Thus the slope of the line from $(0, 0)$ to $(h, f(h))$ oscillates between h (when $f(h) = h^2$ and $\frac{f(h)-0}{h-0} = h$) and $-h$ (when $f(h) = -h^2$ and $\frac{f(h)-0}{h-0} = -h$) as h tends to zero. So, the limit of the slope as h tends to zero is 0, which is the derivative of f at the origin. Another way to see this is to observe that

    $$\lim_{h \to 0} \frac{f(h) - f(0)}{h} = \lim_{h \to 0} \left(\frac{h^2 \sin(\frac{1}{h})}{h} \right)$$
    $$= \lim_{h \to 0} h \sin(\frac{1}{h})$$
    $$= 0,$$

    since $\lim_{h \to 0} h = 0$ and $-1 \le \sin(\frac{1}{h}) \le 1$ for any h. Thus f is differentiable at $x = 0$, and $f'(0) = 0$.

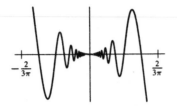

Figure 2.8

9.

(a) The graph of $g(r)$ does not have a break or jump at $r = 2$, and so $g(r)$ is continuous there. This is confirmed by the fact that
    $$g(2) = 1 + \cos(\pi 2/2) = 1 + (-1) = 0$$
    so the value of $g(r)$ as you approach $r = 2$ from the left is the same as the value when you approach $r = 2$ from the right.

(b) The graph of $g(r)$ does not have a corner at $r = 2$, even after zooming in, so $g(r)$ appears to be differentiable at $r = 0$. This is confirmed by the fact that $\cos(\pi r/2)$ is at the bottom of a trough at $r = 2$, and so its slope is 0 there. Thus the slope to the left of $r = 2$ is the same as the slope to the right of $r = 2$.

13. Since $f(1) = 1$ and we showed that $f'(1) = 2$, the local linearization is
    $$f(x) \approx 1 + 2(x - 1) = 2x - 1.$$

CHAPTER THREE

Solutions for Section 3.1

1. (a) Lower estimate = $(45)(2) + (16)(2) + (0)(2) = 122$ feet.
        Upper estimate = $(88)(2) + (45)(2) + (16)(2) = 298$ feet.

    (b)

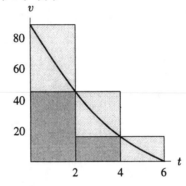

5. Using $\Delta t = 2$,

$$\text{Left-hand sum} = v(0) \cdot 2 + v(2) \cdot 2 + v(4) \cdot 2$$
$$= 1(2) + 5(2) + 17(2)$$
$$= 46$$
$$\text{Right-hand sum} = v(2) \cdot 2 + v(4) \cdot 2 + v(6) \cdot 2$$
$$= 5(2) + 17(2) + 37(2)$$
$$= 118$$
$$\text{Average} = \frac{46 + 118}{2} = 82$$
$$\text{Distance traveled} \approx 82 \text{ meters.}$$

9. Just counting the squares (each of which has area 10, in units of meters), and allowing for the broken squares, we can see that the area under the curve from 0 to 6 is between 140 and 150. Hence the distance traveled is between 140 and 150 meters.

Solutions for Section 3.2

1.

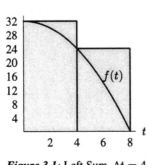

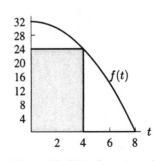

Figure 3.1: Left Sum, $\Delta t = 4$ **Figure 3.2**: Right Sum, $\Delta t = 4$

    (a) Left-hand sum = $32(4) + 24(4) = 224$.
    (b) Right-hand sum = $24(4) + 0(4) = 96$.

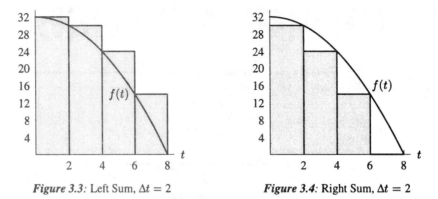

Figure 3.3: Left Sum, $\Delta t = 2$ *Figure 3.4:* Right Sum, $\Delta t = 2$

(c)   Left-hand sum $= 32(2) + 30(2) + 24(2) + 14(2) = 200$.
(d)   Right-hand sum $= 30(2) + 24(2) + 14(2) + 0(2) = 136$.

5.

n	2	10	50	250
Left-hand Sum	-0.394991	-0.0920539	-0.0429983	-0.0335556
Right-hand Sum	0.189470	0.0248382	-0.0196199	-0.0288799

There is no obvious guess as to what the limiting sum is. Moreover, since $\sin(t^2)$ is *not* monotonic on $[2, 3]$, we cannot be sure that the true value is between -0.0335556 and -0.0288799.

9.   Left-hand sum gives: $1^2(1/4) + (1.25)^2(1/4) + (1.5)^2(1/4) + (1.75)^2(1/4) = 1.96875$.
Right-hand sum gives: $(1.25)^2(1/4) + (1.5)^2(1/4) + (1.75)^2(1/4) + (2)^2(1/4) = 2.71875$.

We estimate the value of the integral by taking the average of these two sums, which is 2.34375. Since x^2 is monotonic on $1 \leq x \leq 2$, the true value of the integral lies between 1.96875 and 2.71875. Thus the most our estimate could be off is 0.375. We expect it to be much closer. (And it is—the true value of the integral is $7/3 \approx 2.333$.)

13.   The graph of $y = 7 - x^2$ has intercepts $x = \pm\sqrt{7}$. See Figure 3.5. Therefore we have

$$\text{Area} = \int_{-\sqrt{7}}^{\sqrt{7}} (7 - x^2)\, dx = 24.7.$$

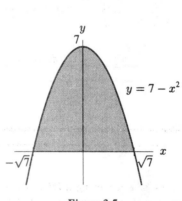

Figure 3.5

17.   Since $x^{1/2} \leq x^{1/3}$ for $0 \leq x \leq 1$, we have

$$\text{Area} = \int_0^1 (x^{1/3} - x^{1/2})\, dx = 0.0833.$$

21.

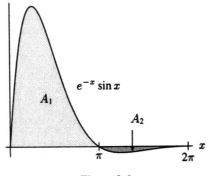

Figure 3.6

Looking at the graph of $e^{-x} \sin x$ for $0 \leq x \leq 2\pi$ in Figure 3.6, we see that the area, A_1, below the curve for $0 \leq x \leq \pi$ is much greater than the area, A_2, above the curve for $\pi \leq x \leq 2\pi$. Thus, the integral is

$$\int_0^{2\pi} e^{-x} \sin x \, dx = A_1 - A_2 > 0.$$

25. (a) We have $\Delta x = (1 - 0)/n = 1/n$ and $x_i = 0 + i \cdot \Delta x = i/n$. So we get

$$\text{Right-hand sum} = \sum_{i=1}^{n} (x_i)^4 \Delta x = \sum_{i=1}^{n} \left(\frac{i}{n}\right)^4 \left(\frac{1}{n}\right) = \sum_{i=1}^{n} \frac{i^4}{n^5}.$$

(b) The CAS gives

$$\text{Right-hand sum} = \sum_{i=1}^{n} \frac{i^4}{n^5} = \frac{6n^4 + 15n^3 + 10n^2 - 1}{30n^4}.$$

(The results may look slightly different depending on the CAS you use.)

(c) Using a CAS or by hand, we get

$$\lim_{n \to \infty} \frac{6n^4 + 15n^3 + 10n^2 - 1}{30n^4} = \lim_{n \to \infty} \frac{6n^4}{30n^4} = \frac{1}{5}.$$

The numerator is dominated by the highest power term, which is $6n^4$, so when n is large, the ratio behaves like $6n^4/30n^4 = 1/5$ as $n \to \infty$. Thus we see that

$$\int_0^1 x^4 dx = \frac{1}{5}.$$

Solutions for Section 3.3

1. (a) One small box on the graph corresponds to moving at 750 ft/min for 15 seconds, which corresponds to a distance of 187.5 ft. Estimating the area beneath the velocity curves, we find:
Distance traveled by car 1 $\approx$ 5.5 boxes = 1031.25 ft.
Distance traveled by car 2 $\approx$ 3 boxes = 562.5 ft.

(b) The two cars will have gone the same distance when the areas beneath their velocity curves are equal. Since the two areas overlap, they are equal when the two shaded regions have equal areas, at $t \approx 1.6$ minutes. See Figure 3.7.

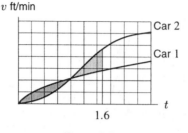

Figure 3.7

5. The units of measurement are dollars.

9. Sketch the graph of f on $1 \leq x \leq 3$. The integral is the area under the curve, which is a trapezoidal area. So the average value is

$$\frac{1}{3-1} \int_1^3 (4x + 7)\, dx = \frac{1}{2} \cdot \frac{11+19}{2} \cdot 2 = \frac{30}{2} = 15.$$

13. The time period 9am to 5pm is represented by the time $t = 0$ to $t = 8$ and $t = 24$ to $t = 32$. The area under the curve, or total number of worker-hours for these times, is about 9 boxes or $9(80) = 720$ worker-hours. The total cost for 9am to 5pm is $(720)(10) = \$7200$. The area under the rest of the curve is about 5.5 boxes, or $5.5(80) = 440$ worker-hours. The total cost for this time period is $(440)(15) = \$6600$. The total cost is about $7200 + 6600 = \$13,800$.

17. (a)   Average value $= \displaystyle\int_0^1 \sqrt{1 - x^2}\, dx = 0.79$

    (b)   The area between the graph of $y = 1 - x$ and the x-axis is 0.5. Because the graph of $y = \sqrt{1 - x^2}$ is concave down, it lies above the line $y = 1 - x$, so its average value is above 0.5. See figure below.

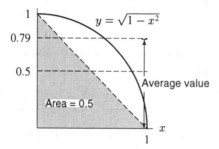

21. (a)   Since $t = 0$ to $t = 31$ covers January:

    $$\left(\begin{array}{c} \text{Average number of} \\ \text{daylight hours in January} \end{array} \right) = \frac{1}{31} \int_0^{31} \left[12 + 2.4 \sin(0.0172(t - 80)) \right]\, dt.$$

    Using left and right sums with $n = 100$ gives

    $$\text{Average} \approx \frac{306}{31} \approx 9.9 \text{ hours.}$$

    (b)   Assuming it is not a leap year, the last day of May is $t = 151 (= 31 + 28 + 31 + 30 + 31)$ and the last day of June is $t = 181 (= 151 + 30)$. Again finding the integral numerically:

    $$\left(\begin{array}{c} \text{Average number of} \\ \text{daylight hours in June} \end{array} \right) = \frac{1}{30} \int_{151}^{181} \left[12 + 2.4 \sin(0.0172(t - 80)) \right]\, dt$$

    $$\approx \frac{431}{30} \approx 14.4 \text{ hours.}$$

(c)

$$\text{(Average for whole year)} = \frac{1}{365}\int_0^{365}\left[12 + 2.4\sin(0.0172(t-80))\right]\,dt$$

$$\approx \frac{4381}{365} \approx 12.0 \text{ hours.}$$

(d) The average over the whole year should be 12 hours, as computed in (c). Since Madrid is in the northern hemisphere, the average for a winter month, such as January, should be less than 12 hours (it is 9.9 hours) and the average for a summer month, such as June, should be more than 12 hours (it is 14.4 hours).

25. We'll show that in terms of the average value of f,

$$\mathrm{I} > \mathrm{II} = \mathrm{IV} > \mathrm{III}$$

Using part (a) of the solution to Problem 24 on page 179 of the text, we have

$$\begin{aligned}\text{Average value} \atop \text{of } f \text{ on II} &= \frac{\int_0^2 f(x)\,dx}{2} = \frac{\frac{1}{2}\int_{-2}^2 f(x)\,dx}{2}\\ &= \frac{\int_{-2}^2 f(x)\,dx}{4}\\ &= \text{Average value of } f \text{ on IV.}\end{aligned}$$

Since f is decreasing on $[0, 5]$, the average value of f on the interval $[0, c]$ is decreasing as a function of c. The larger the interval the more low values of f are included. Hence

$$\text{Average value of } f \atop \text{on } [0,1] > \text{Average value of } f \atop \text{on } [0,2] > \text{Average value of } f \atop \text{on } [0,5]$$

Solutions for Section 3.4

1. We find the changes in $f(x)$ between any two values of x by counting the area between the curve of $f'(x)$ and the x-axis. Since $f'(x)$ is linear throughout, this is quite easy to do. From $x = 0$ to $x = 1$, we see that $f'(x)$ outlines a triangle of area $1/2$ below the x-axis (the base is 1 and the height is 1). By the Fundamental Theorem,

$$\int_0^1 f'(x)\,dx = f(1) - f(0),$$

so

$$f(0) + \int_0^1 f'(x)\,dx = f(1)$$

$$f(1) = 2 - \frac{1}{2} = \frac{3}{2}$$

Similarly, between $x = 1$ and $x = 3$ we can see that $f'(x)$ outlines a rectangle below the x-axis with area -1, so $f(2) = 3/2 - 1 = 1/2$. Continuing with this procedure (note that at $x = 4$, $f'(x)$ becomes positive), we get the table below.

x	0	1	2	3	4	5	6
$f(x)$	2	3/2	1/2	$-1/2$	-1	$-1/2$	1/2

5. First rewrite each of the quantities in terms of f', since we have the graph of f'. If A_1 and A_2 are the positive areas shown in Figure 3.8:

$$f(3) - f(2) = \int_2^3 f'(t)\,dt = -A_1$$

$$f(4) - f(3) = \int_3^4 f'(t)\,dt = -A_2$$

$$\frac{f(4) - f(2)}{2} = \frac{1}{2}\int_2^4 f'(t)\,dt = -\frac{A_1 + A_2}{2}$$

Since Area $A_1 >$ Area A_2,

$$A_2 < \frac{A_1 + A_2}{2} < A_1$$

so

$$-A_1 < -\frac{A_1 + A_2}{2} < -A_2$$

and therefore

$$f(3) - f(2) < \frac{f(4) - f(2)}{2} < f(4) - f(3).$$

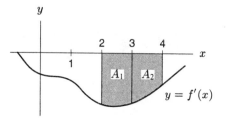

Figure 3.8

9. (a) Over the interval $[-1, 3]$, we estimate that the total change of the population is about 1.5, by counting boxes between the curve and the x-axis; we count about 1.5 boxes below the x-axis from $x = -1$ to $x = 1$ and about 3 above from $x = 1$ to $x = 3$. So the average rate of change is just the total change divided by the length of the interval, that is $1.5/4 = 0.375$ thousand/hour.

   (b) We can estimate the total change of the algae population by counting boxes between the curve and the x-axis. Here, there is about 1 box above the x-axis from $x = -3$ to $x = -2$, about 0.75 of a box below the x-axis from $x = -2$ to $x = -1$, and a total change of about 1.5 boxes thereafter (as discussed in part (a)). So the total change is about $1 - 0.75 + 1.5 = 1.75$ thousands of algae.

13.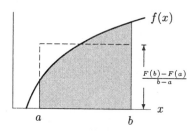

Note that we are using the interpretation of the definite integral as the length of the interval times the average value of the function on that interval, which we developed in Section 3.3.

17. Note that $\int_a^b f(z)\,dz = \int_a^b f(x)\,dx$. Thus, we have

$$\int_a^b cf(z)\,dz = c\int_a^b f(z)\,dz = 8c.$$

21.   (a) $\int_{-1}^{1} e^{x^2}\, dx > 0$, since $e^{x^2} > 0$, and $\int_{-1}^{1} e^{x^2}\, dx$ represents the area below the curve $y = e^{x^2}$.

(b)   Looking at the figure below, we see that $\int_{0}^{1} e^{x^2}\, dx$ represents the area under the curve. This area is clearly greater than zero, but it is less than e since it fits inside a rectangle of width 1 and height e (with room to spare). Thus

$$0 < \int_{0}^{1} e^{x^2}\, dx < e < 3.$$

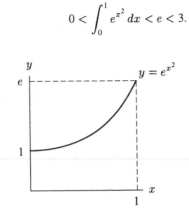

25.   We know that the average value of $v(x) = 4$, so

$$\frac{1}{6-1}\int_{1}^{6} v(x)\, dx = 4, \quad \text{and thus} \quad \int_{1}^{6} v(x)\, dx = 20.$$

Similarly, we are told that

$$\frac{1}{8-6}\int_{6}^{8} v(x)\, dx = 5, \quad \text{so} \quad \int_{6}^{8} v(x)\, dx = 10.$$

The average value for $1 \le x \le 8$ is given by

$$\text{Average value} = \frac{1}{8-1}\int_{1}^{8} v(x)\, dx = \frac{1}{7}\left(\int_{1}^{6} v(x)\, dx + \int_{6}^{8} v(x)\, dx\right) = \frac{20+10}{7} = \frac{30}{7}.$$

Solutions for Chapter 3 Review

1.   (a)   We calculate the right- and left-hand sums as follows:

$$\text{Left} = 2[80 + 52 + 28 + 10] = 340 \text{ ft.}$$
$$\text{Right} = 2[52 + 28 + 10 + 0] = 180 \text{ ft.}$$

Our best estimate will be the average of these two sums,

$$\text{Best} = \frac{\text{Left} + \text{Right}}{2} = \frac{340 + 180}{2} = 260 \text{ ft.}$$

(b)   Since v is decreasing throughout,

$$\text{Left} - \text{Right} = \Delta t \cdot [f(0) - f(8)]$$
$$= 80\Delta t.$$

Since our best estimate is the average of Left and Right, the maximum error is $(80)\Delta t/2$. For $(80)\Delta t/2 \le 20$, we must have $\Delta t \le 1/2$. In other words, we must measure the velocity every 0.5 second.

5.   Distance traveled = $\int_0^{1.1} \sin(t^2)\, dt \approx 0.40.$

9.   Since the θ intercepts of $y = \sin\theta$ are
$\theta = 0, \pi, 2\pi, \ldots,$

Area $= \int_0^{\pi} 1\, d\theta - \int_0^{\pi} \sin\theta\, d\theta = \pi - 2 \approx 1.14.$

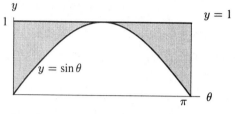

13.

By the FTC, we know that $\int_a^b f(x)\, dx = F(b) - F(a).$

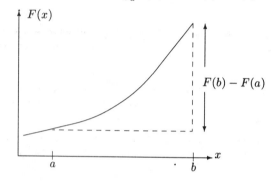

17.   (a)

$$\text{Average population} \ = \ \frac{1}{10}\int_0^{10} 67.38(1.026)^t\, dt$$

Evaluating the integral numerically gives

$$\text{Average population} \ \approx \ 76.8 \text{ million}$$

(b)   In 1980, $t = 0$, and $P = 67.38(1.026)^0 = 67.38.$
In 1990, $t = 10$, and $P = 67.38(1.026)^{10} = 87.10.$
Average$= \frac{1}{2}(67.38 + 87.10) = 77.24$ million.

(c)   If P had been linear, the average value found in (a) would have been the one we found in (b). Since the population graph is concave up, it is below the secant line. Thus, the actual values of P are less than the corresponding values on the secant line, and so the average found in (a) is smaller than that in (b).

21.   (a)   About 300 meter^3/sec.
(b)   About 250 meter^3/sec.
(c)   Looking at the graph, we can see that the 1996 flood reached its maximum just between March and April, for a high of about 1250 meter^3/sec. Similarly, the 1957 flood reached its maximum in mid-June, for a maximum flow rate of 3500 meter^3/sec.
(d)   The 1996 flood lasted about 1/3 of a month, or about 10 days. The 1957 flood lasted about 4 months.
(e)   The area under the controlled flood graph is about 2/3 box. Each box represents 500 meter^3/sec for one month. Since

$$1 \text{ month} = 30\frac{\text{days}}{\text{month}} \cdot 24\frac{\text{hours}}{\text{day}} \cdot 60\frac{\text{minutes}}{\text{hour}} \cdot 60\frac{\text{seconds}}{\text{minute}}$$
$$= 2.592 \cdot 10^6 \approx 3 \cdot 10^6 \text{seconds},$$

each box represents

$$\text{Flow} \approx (500 \text{ meter}^3/\text{sec}) \cdot (2.6 \cdot 10^6 \text{ sec}) = 13 \cdot 10^8 \text{ meter}^3 \text{of water.}$$

So, for the artificial flood,

$$\text{Additional flow} \approx \frac{2}{3} \cdot 13 \cdot 10^8 = 9 \cdot 10^8 \text{ meter}^3 \approx 10^9 \text{ meter}^3.$$

(f) The 1957 flood released a volume of water represented by about 12 boxes above the 250 meter/sec baseline. Thus, for the natural flood,

$$\text{Additional flow} \approx 12 \cdot 15 \cdot 10^8 = 1.8 \cdot 10^{10} \approx 2 \cdot 10^{10} \text{ meter}^3.$$

So, the natural flood was nearly 20 times larger than the controlled flood and lasted much longer.

25. (a) We know that $\int_2^5 f(x)\,dx = \int_0^5 f(x)\,dx - \int_0^2 f(x)\,dx$. By symmetry, $\int_0^2 f(x)\,dx = \frac{1}{2}\int_{-2}^2 f(x)\,dx$, so $\int_2^5 f(x)\,dx = \int_0^5 f(x)\,dx - \frac{1}{2}\int_{-2}^2 f(x)\,dx$.

(b) $\int_2^5 f(x)\,dx = \int_{-2}^5 f(x)\,dx - \int_{-2}^2 f(x)\,dx = \int_{-2}^5 f(x)\,dx - 2\int_{-2}^0 f(x)\,dx.$

(c) Using symmetry again, $\int_0^2 f(x)\,dx = \frac{1}{2}(\int_{-2}^5 f(x)\,dx - \int_2^5 f(x)\,dx).$

Solutions to Problems on the Definite Integral

1. The statement is rarely true. The graph of almost any non-linear monotonic function, such as x^{10} for $0 < x < 1$, should provide convincing geometric evidence. Furthermore, if the statement were true, then (LHS+RHS)/2 would always give the exact value of the definite integral. This is not true.

5. Since the integrand is increasing on $[1, 2]$, the left-hand sum is the lower sum and the right-hand sum is the upper sum. For $n = 30$, LHS ≈ 2.852 (rounding down) and RHS ≈ 2.919 (rounding up). Since the left and right sums differ by 0.067, their average must be within 0.0335 of the true value, so $\int_1^2 2^x\,dx = 2.886$ to the required accuracy.

9. Since the integrand is increasing on $[-2, -1]$, the left-hand sum is the lower sum and the right-hand sum is the upper sum. For $n = 10$, LHS ≈ 0.0045 (rounding down) and RHS ≈ 0.0276 (rounding up). Since the left and right sums differ by 0.0231, their average must be within 0.01155 of the true value, so $\int_{-2}^{-1} \cos^3 y\,dy = 0.016$ to the required accuracy.

13. Let $a = x_0 < x_1 < x_2 < \cdots < x_{n-1} < x_n$ be the endpoints of the subdivision. The lower sum is

$$\sum_{i=1}^n m_i \Delta x_i$$

and the upper sum is

$$\sum_{i=1}^n M_i \Delta x_i,$$

where m_i is the greatest lower bound for f on $[x_{i-1}, x_i]$ and M_i is the least upper bound for f on the same interval. Since a lower bound for a set must be less than or equal to an upper bound for the same set, $m_i \leq M_i$. Thus the lower sum is less than or equal to the upper sum.

17. Let A be a lower sum and let B be an upper sum. Using Problem 16, choose a subdivision which is a refinement of both the subdivisions used in these sums, and let A' and B' be the lower and upper sums corresponding to this subdivision. By Problem 14, $A \leq A'$, and by Problem 15, $B' \leq B$. Finally, by Problem 13, $A' \leq B'$. So

$$A \leq A' \leq B' \leq B,$$

hence $A \leq B$.

CHAPTER FOUR

1. The derivative, $f'(x)$, is defined as
$$f'(x) = \lim_{h \to 0} \frac{f(x+h) - f(x)}{h}.$$

   If $f(x) = 7$, then
$$f'(x) = \lim_{h \to 0} \frac{7 - 7}{h} = \lim_{h \to 0} \frac{0}{h} = 0.$$

5. $y' = 12x^{11}$.

9. $y' = \frac{4}{3}x^{1/3}$.

13. $f'(x) = \frac{1}{4}x^{-3/4}$.

17. $y' = -12x^3 - 12x^2 - 6$.

21. $y = x + \frac{1}{x}$, so $y' = 1 - \frac{1}{x^2}$.

25. The functions whose derivatives don't exist at $x = 0$ are in problems 8, 10, 11, 12, 13, and 15.

29. $y' = -\frac{2}{3z^3}$.   (power rule and sum rule)

33. $g'(x) = \dfrac{12}{\sqrt[6]{x^5}} - \dfrac{18}{\sqrt[3]{x^5}}$, using the power and sum rules.

37. Decreasing means $f'(x) < 0$:
$$f'(x) = 4x^2(x - 3) < 0,$$

   so $x < 3$ and $x \neq 0$. Concave up means $f''(x) > 0$:
$$f''(0) = 12x^2 - 24x > 0$$
$$12x(x - 2) > 0$$
$$x < 0 \quad \text{or} \quad x > 2.$$

   So, both conditions hold for $x < 0$ or $2 < x < 3$.

41.
$$f'(x) = 6x^2 - 4x \quad \text{so} \quad f'(1) = 6 - 4 = 2.$$
   Thus the equation of the tangent line is $(y - 1) = 2(x - 1)$ or $y = 2x - 1$.

45. The slope of the tangent lines to $y = x^2 - 2x + 4$ is $y' = 2x - 2$. For a line through the origin, $y = mx$. So, at the tangent point, $x^2 - 2x + 4 = mx$ where $m = y' = 2x - 2$.
$$x^2 - 2x + 4 = (2x - 2)x$$
$$x^2 - 2x + 4 = 2x^2 - 2x$$
$$-x^2 + 4 = 0$$
$$-(x + 2)(x - 2) = 0$$
$$x = 2, -2.$$

   Thus, the points of tangency are $(2, 4)$ and $(-2, 12)$. The lines through these points and the origin are $y = 2x$ and $y = -6x$, respectively. Graphically, this can be seen in Figure 4.1:

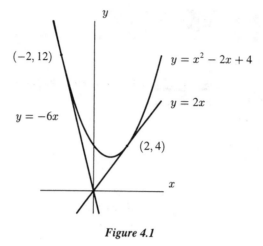

Figure 4.1

49. $f(x) = \frac{1}{x} = x^{-1}$ $f'(x) = -x^{-2} = -\frac{1}{x^2}$.

The tangent line at $x = 1$ will have slope $f'(1) = -1$. See Figure 4.2. Using the point $(1, 1)$ which lies on the line, we obtain the equation $y = -x + 2$. We approximate $f(2)$ by using the y-value corresponding to $x = 2$, so $f(2) \approx 0$.

Similarly, the tangent line at $x = 100$ will have slope $f'(100) = \frac{-1}{(100)^2} = -0.0001$. The equation of the line is then $y = -0.0001x + 0.02$. The approximate value of $f(2)$ predicted by this tangent line is $f(2) \approx 0.0198$.

The actual value of $f(2)$ is $\frac{1}{2}$, so the approximation from $x = 100$ is better than that from $x = 1$. This is because the slope changes less between $x = 2$ and $x = 100$ than it does between $x = 1$ and $x = 2$.

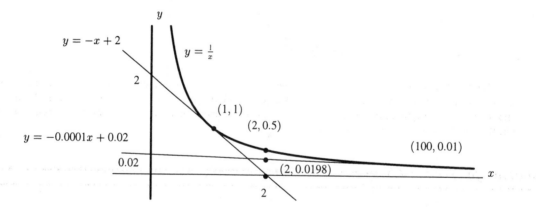

Figure 4.2

Solutions for Section 4.2

1. $f'(x) = 2e^x + 2x$.

5. $y' = 10x + (\ln 2)2^x$.

9. $\frac{dy}{dx} = \frac{1}{3}(\ln 3)3^x - \frac{33}{2}(x^{-\frac{3}{2}})$.

13. $y = e^\theta e^{-1}$ $y' = \dfrac{d}{d\theta}(e^\theta e^{-1}) = e^{-1}\dfrac{d}{d\theta}e^\theta = e^\theta e^{-1} = e^{\theta-1}$.

17. $f'(t) = (\ln(\ln 3))(\ln 3)^t$.

21. $f'(x) = \pi^2 x^{(\pi^2-1)} + (\pi^2)^x \ln(\pi^2)$

25. Once again, this is a product of two functions, 2^x and $\frac{1}{x}$, each of which we can take the derivative of; but we don't know how to take the derivative of the product.

29. $f'(z) = (\ln\sqrt{4})(\sqrt{4})^z = (\ln 2)2^z$.

33. $\dfrac{dV}{dt} = 75(1.35)^t \ln 1.35 = 22.5(1.35)^t$.

37.

$$g(x) = ax^2 + bx + c \qquad\qquad f(x) = e^x$$
$$g'(x) = 2ax + b \qquad\qquad\quad f'(x) = e^x$$
$$g''(x) = 2a \qquad\qquad\qquad\quad f''(x) = e^x$$

So, using $g''(0) = f''(0)$, etc., we have $2a = 1$, $b = 1$, and $c = 1$, and thus $g(x) = \frac{1}{2}x^2 + x + 1$, as shown in the figure below.

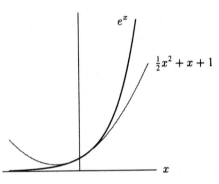

The two functions do look very much alike near $x = 0$. They both increase for large values of x, but e^x increases much more quickly. For very negative values of x, the quadratic goes to ∞ whereas the exponential goes to 0. By choosing a function whose first few derivatives agreed with the exponential when $x = 0$, we got a function which looks like the exponential for x-values near 0.

Solutions for Section 4.3

1. By the product rule, $f'(x) = 2x(x^3 + 5) + x^2(3x^2) = 2x^4 + 3x^4 + 10x = 5x^4 + 10x$. Alternatively, $f'(x) = (x^5 + 5x^2)' = 5x^4 + 10x$. The two answers should, and do, match.

5. $y' = \frac{1}{2\sqrt{x}}2^x + \sqrt{x}(\ln 2)2^x$.

9. $w' = (3t^2 + 5)(t^2 - 7t + 2) + (t^3 + 5t)(2t - 7)$.

13. $g'(w) = \dfrac{3.2w^{2.2}(5^w) - (\ln 5)(w^{3.2})5^w}{5^{2w}} = \dfrac{3.2w^{2.2} - w^{3.2}(\ln 5)}{5^w}$.

17.
$$f'(x) = \frac{x(2x) - (x^2 + 3)(1)}{x^2}$$
$$= \frac{2x^2 - x^2 - 3}{x^2}$$
$$= \frac{x^2 - 3}{x^2}.$$

21. $w'(x) = \dfrac{17e^x(2^x) - (\ln 2)(17e^x)2^x}{2^{2x}} = \dfrac{17e^x(2^x)(1 - \ln 2)}{2^{2x}} = \dfrac{17e^x(1 - \ln 2)}{2^x}$.

25. Using the product rule, we have

$$f'(x) = e^{-x} - xe^{-x}$$
$$f''(x) = -e^{-x} - e^{-x} + xe^{-x} = e^{-x}(x - 2).$$

Since $e^{-x} > 0$, for all x, we have $f''(x) < 0$ if $x - 2 < 0$, that is, $x < 2$.

29.

$$f(x) = e^x e^{2x}$$
$$f'(x) = e^x(e^{2x})' + (e^x)'e^{2x}$$
$$= 2e^x e^{2x} + e^x e^{2x} \text{ (from Problem 28)}$$
$$= 3e^{3x}.$$

33. Since

$$x^{1/2} \cdot x^{1/2} = x,$$

we differentiate to obtain

$$\frac{d}{dx}(x^{1/2}) \cdot x^{1/2} + x^{1/2} \cdot \frac{d}{dx}(x^{1/2}) = 1.$$

Now solve for $d(x^{1/2})/dx$:

$$2x^{1/2}\frac{d}{dx}(x^{1/2}) = 1$$
$$\frac{d}{dx}(x^{1/2}) = \frac{1}{2x^{1/2}}.$$

37. (a) $f(140) = 15{,}000$ says that 15,000 skateboards are sold when the cost is \$140 per board.

$f'(140) = -100$ means that if the price is increased from \$140, roughly speaking, every dollar of increase will decrease the total sales by 100 boards.

(b) $\dfrac{dR}{dp} = \dfrac{d}{dp}(p \cdot q) = \dfrac{d}{dp}\left(p \cdot f(p)\right) = f(p) + pf'(p).$

So,

$$\frac{dR}{dp}\bigg|_{p=140} = f(140) + 140f'(140)$$
$$= 15{,}000 + 140(-100) = 1000.$$

(c) From (b) we see that $\dfrac{dR}{dp}\bigg|_{p=140} = 1000 > 0$. This means that the revenue will increase by about \$1000 if the price is raised by \$1.

41. Assume for $g(x) \neq f(x)$, $g'(x) = g(x)$ and $g(0) = 1$. Then for

$$h(x) = \frac{g(x)}{e^x}$$
$$h'(x) = \frac{g'(x)e^x - g(x)e^x}{(e^x)^2} = \frac{e^x(g'(x) - g(x))}{(e^x)^2} = \frac{g'(x) - g(x)}{e^x}.$$

But, since $g(x) = g'(x)$, $h'(x) = 0$, so $h(x)$ is constant. Thus, the ratio of $g(x)$ to e^x is constant. Since $\dfrac{g(0)}{e^0} = \dfrac{1}{1} = 1$, $\dfrac{g(x)}{e^x}$ must equal 1 for all x. Thus $g(x) = e^x = f(x)$ for all x, so f and g are the same function.

Solutions for Section 4.4

1. $f'(x) = 99(x + 1)^{98} \cdot 1 = 99(x + 1)^{98}$.

5. $w' = 100(\sqrt{t} + 1)^{99}\left(\frac{1}{2\sqrt{t}}\right) = \frac{50}{\sqrt{t}}(\sqrt{t} + 1)^{99}$.

9. $k'(x) = 4(x^3 + e^x)^3(3x^2 + e^x)$.

13. $y' = \frac{3}{2}e^{\frac{3}{2}w}$.

17. $y' = 1 \cdot e^{-t^2} + te^{-t^2}(-2t)$

21. $f'(t) = 1 \cdot e^{5-2t} + te^{5-2t}(-2) = e^{5-2t}(1 - 2t)$

25.

$$f'(w) = (e^{w^2})(10w) + (5w^2 + 3)(e^{w^2})(2w)$$
$$= 2we^{w^2}(5 + 5w^2 + 3)$$
$$= 2we^{w^2}(5w^2 + 8).$$

29. $f'(y) = e^{e^{(y^2)}}\left[(e^{y^2})(2y)\right] = 2ye^{[e^{(y^2)}+y^2]}$.

33.

$$f'(x) = [10(2x + 1)^9(2)][(3x - 1)^7] + [(2x + 1)^{10}][7(3x - 1)^6(3)]$$
$$= (2x + 1)^9(3x - 1)^6[20(3x - 1) + 21(2x + 1)]$$
$$= [(2x + 1)^9(3x - 1)^6](102x + 1)$$
$$f''(x) = \quad [9(2x + 1)^8(2)(3x - 1)^6 + (2x + 1)^9(6)(3x - 1)^5(3)](102x + 1)$$
$$+ (2x + 1)^9(3x - 1)^6(102).$$

37. (a) Differentiating $g(x) = \sqrt{f(x)} = (f(x))^{1/2}$, we have

$$g'(x) = \frac{1}{2}(f(x))^{-1/2} \cdot f'(x) = \frac{f'(x)}{2\sqrt{f(x)}}$$
$$g'(1) = \frac{f'(1)}{2\sqrt{f(1)}} = \frac{3}{2\sqrt{4}} = \frac{3}{4}.$$

   (b) Differentiating $h(x) = f(\sqrt{x})$, we have

$$h'(x) = f'(\sqrt{x}) \cdot \frac{1}{2\sqrt{x}}$$
$$h'(1) = f'(\sqrt{1}) \cdot \frac{1}{2\sqrt{1}} = \frac{f'(1)}{2} = \frac{3}{2}.$$

41. (a)

$$\frac{dQ}{dt} = \frac{d}{dt}e^{-0.000121t}$$
$$= -0.000121e^{-0.000121t}$$

   (b)

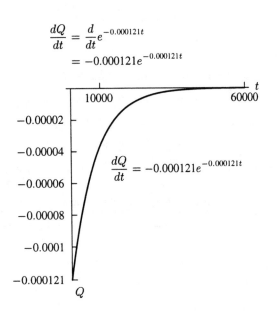

45. (a)

$$\frac{dm}{dv} = \frac{d}{dv}\left[m_0\left(1 - \frac{v^2}{c^2}\right)^{-1/2}\right]$$

$$= m_0\left(-\frac{1}{2}\right)\left(1 - \frac{v^2}{c^2}\right)^{-3/2}\left(-\frac{2v}{c^2}\right)$$

$$= \frac{m_0 v}{c^2}\frac{1}{\sqrt{\left(1 - \frac{v^2}{c^2}\right)^3}}.$$

(b) $\dfrac{dm}{dv}$ represents the rate of change of mass with respect to the speed v.

Solutions for Section 4.5

1.

TABLE 4.1

x	$\cos x$	Difference Quotient	$-\sin x$
0	1.0	-0.0005	0.0
0.1	0.995	-0.10033	-0.099833
0.2	0.98007	-0.19916	-0.19867
0.3	0.95534	-0.296	-0.29552
0.4	0.92106	-0.38988	-0.38942
0.5	0.87758	-0.47986	-0.47943
0.6	0.82534	-0.56506	-0.56464

5. $z' = -4\sin(4\theta)$.

9. $f'(x) = (e^{\cos x})(-\sin x) = -\sin x e^{\cos x}$.

13. $f'(x) = \dfrac{\cos x}{\cos^2(\sin x)}$.

17. $z' = e^{\cos\theta} - \theta(\sin\theta)e^{\cos\theta}$.

21. $f'(x) = (e^{-2x})(-2)(\sin x) + (e^{-2x})(\cos x) = -2\sin x(e^{-2x}) + (e^{-2x})(\cos x) = e^{-2x}[\cos x - 2\sin x]$.

25. $z' = \dfrac{-3e^{-3\theta}}{\cos^2(e^{-3\theta})}$.

29. $f'(\theta) = 2\theta\sin\theta + \theta^2\cos\theta + 2\cos\theta - 2\theta\sin\theta - 2\cos\theta = \theta^2\cos\theta$.

33.   (a) $\dfrac{dy}{dt} = -\dfrac{4.9\pi}{6}\sin\left(\dfrac{\pi}{6}t\right)$. It represents the rate of change of the depth of the water.

(b) $\dfrac{dy}{dt}$ is zero where the tangent line to the curve $y(t)$ is horizontal. $\dfrac{dy}{dt} = 0$ occurs when $\sin(\frac{\pi}{6}t) = 0$, or at $t =$ 6 am, 12 noon, 6 pm and 12 midnight. When $\dfrac{dy}{dt} = 0$, the depth of the water is no longer changing. Therefore, it has either just finished rising or just finished falling, and we know that the harbor's level is at a maximum or a minimum.

37.   Using the triangle OSL in the figure below, we label the distance x.

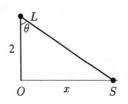

We want to calculate $dx/d\theta$. First we must find x as a function of θ. From the triangle, we see

$$\frac{x}{2} = \tan\theta \quad \text{so} \quad x = 2\tan\theta.$$

Thus,

$$\frac{dx}{d\theta} = \frac{2}{\cos^2\theta}.$$

Solutions for Section 4.6

1. $f'(t) = \frac{2t}{t^2+1}$.

5. $f'(z) = -1(\ln z)^{-2} \cdot \frac{1}{z} = \frac{-1}{z(\ln z)^2}$.

9. $f'(x) = \frac{1}{e^x+1} \cdot e^x$.

13. $f'(w) = \frac{1}{\cos(w-1)}[-\sin(w-1)] = -\tan(w-1)$.

    [This could be done easily using the answer from Problem 6 and the chain rule.]

17. $g(\alpha) = \alpha$, so $g'(\alpha) = 1$.

21. $h'(w) = \arcsin w + \dfrac{w}{\sqrt{1-w^2}}$.

25. Let

$$g(x) = \log x.$$

Then

$$10^{g(x)} = x.$$

Differentiating,

$$(\ln 10)[10^{g(x)}]g'(x) = 1$$
$$g'(x) = \frac{1}{(\ln 10)[10^{g(x)}]}$$
$$g'(x) = \frac{1}{(\ln 10)x}.$$

29. (a)

$$f'(x) = \frac{1}{1+x^2} + \frac{1}{1+\frac{1}{x^2}} \cdot \left(-\frac{1}{x^2}\right)$$
$$= \frac{1}{1+x^2} + \left(-\frac{1}{x^2+1}\right)$$
$$= \frac{1}{1+x^2} - \frac{1}{1+x^2}$$
$$= 0$$

    (b) f is a constant function. Checking at a few values of x,

TABLE 4.2

x	$\arctan x$	$\arctan x^{-1}$	$f(x) = \arctan x + \arctan x^{-1}$
1	0.785392	0.7853982	1.5707963
2	1.1071487	0.4636476	1.5707963
3	1.2490458	0.3217506	1.5707963

33.   (a)   Using Pythagoras' theorem, we see

$$z^2 = 0.5^2 + x^2$$

so

$$z = \sqrt{0.25 + x^2}.$$

(b)   We want to calculate dz/dt. Using the chain rule, we have

$$\frac{dz}{dt} = \frac{dz}{dx} \cdot \frac{dx}{dt} = \frac{2x}{2\sqrt{0.25 + x^2}} \frac{dx}{dt}.$$

Because the train is moving at 0.8 km/hr, we know that

$$\frac{dx}{dt} = 0.8 \, \text{km/hr}.$$

At the moment we are interested in $z = 1$ km so

$$1^2 = 0.25 + x^2$$

giving

$$x = \sqrt{0.75} = 0.866 \, \text{km}.$$

Therefore

$$\frac{dz}{dt} = \frac{2(0.866)}{2\sqrt{0.25 + 0.75}} \cdot 0.8 = 0.866 \cdot 0.8 = 0.693 \, \text{km/min}.$$

(c)   We want to know $d\theta/dt$, where θ is as shown in Figure 4.3. Since

$$\frac{x}{0.5} = \tan\theta$$

we know

$$\theta = \arctan\left(\frac{x}{0.5}\right),$$

so

$$\frac{d\theta}{dt} = \frac{1}{1 + (x/0.5)^2} \cdot \frac{1}{0.5} \frac{dx}{dt}.$$

We know that $dx/dt = 0.8$ km/min and, at the moment we are interested in, $x = \sqrt{0.75}$. Substituting gives

$$\frac{d\theta}{dt} = \frac{1}{1 + 0.75/0.25} \cdot \frac{1}{0.5} \cdot 0.8 = 0.4 \, \text{radians/min}.$$

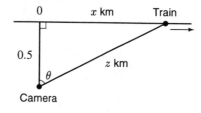

Figure 4.3

Solutions for Section 4.7

1.   We differentiate implicitly both sides of the equation with respect to x.

$$2x + 2y\frac{dy}{dx} = 0,$$

$$\frac{dy}{dx} = -\frac{2x}{2y} = -\frac{x}{y}.$$

5. We differentiate implicitly both sides of the equation with respect to x.

$$\ln x + \ln(y^2) = 3$$

$$\frac{1}{x} + \frac{1}{y^2}(2y)\frac{dy}{dx} = 0$$

$$\frac{dy}{dx} = \frac{-1/x}{2y/y^2} = -\frac{y}{2x}.$$

9. We differentiate implicitly both sides of the equation with respect to x.

$$\cos(xy)\left(y + x\frac{dy}{dx}\right) = 2$$

$$y\cos(xy) + x\cos(xy)\frac{dy}{dx} = 2$$

$$\frac{dy}{dx} = \frac{2 - y\cos(xy)}{x\cos(xy)}.$$

13. First, we must find the slope of the tangent, i.e. $\left.\frac{dy}{dx}\right|_{(1,-1)}$. Differentiating implicitly, we have:

$$y^2 + x(2y)\frac{dy}{dx} = 0,$$

$$\frac{dy}{dx} = -\frac{y^2}{2xy} = -\frac{y}{2x}.$$

Substitution yields $\left.\frac{dy}{dx}\right|_{(1,-1)} = -\frac{-1}{2} = \frac{1}{2}$. Using the point-slope formula for a line, we have that the equation for the tangent line is $y + 1 = \frac{1}{2}(x - 1)$ or $y = \frac{1}{2}x - \frac{3}{2}$.

17. $y = x^{\frac{m}{n}}$. Taking n^{th} powers of both sides of this expression yields $(y)^n = (x^{\frac{m}{n}})^n$, or $y^n = x^m$.

$$\frac{d}{dx}(y^n) = \frac{d}{dx}(x^m)$$

$$ny^{n-1}\frac{dy}{dx} = mx^{m-1}$$

$$\frac{dy}{dx} = \frac{m}{n}\frac{x^{m-1}}{y^{n-1}}$$

$$= \frac{m}{n}\frac{x^{m-1}}{(x^{m/n})^{n-1}}$$

$$= \frac{m}{n}\frac{x^{m-1}}{x^{m-\frac{m}{n}}}$$

$$= \frac{m}{n}x^{(m-1)-(m-\frac{m}{n})} = \frac{m}{n}x^{\frac{m}{n}-1}.$$

21. The slope of the tangent to the curve $y = x^2$ at $x = 1$ is 2 so the equation of such a tangent will be of the form $y = 2x + c$. As the tangent must pass through $(1, 1)$, $c = -1$ and so the required tangent is $y = 2x - 1$.

Any circle centered at $(8, 0)$ will be of the form

$$(x - 8)^2 + y^2 = R^2.$$

The slope of this curve at (x, y) is given by implicit differentiation:

$$2(x - 8) + 2yy' = 0$$

or

$$y' = \frac{8 - x}{y}$$

For the tangent to the parabola to be tangential to the circle we need

$$\frac{8-x}{y} = 2$$

so that at the point of contact of the circle and the line the coordinates are given by (x, y) when $y = 4 - x/2$. Substituting into the equation of the tangent line gives $x = 2$ and $y = 3$. From this we conclude that $R^2 = 45$ so that the equation of the circle is

$$(x-8)^2 + y^2 = 45.$$

Solutions for Section 4.8

1. With $f(x) = 1/x$, we see that the tangent line approximation to f near $x = 1$ is

    $$f(x) \approx f(1) + f'(1)(x-1),$$

    which becomes

    $$\frac{1}{x} \approx 1 + f'(1)(x-1).$$

    Since $f'(x) = -1/x^2$, $f'(1) = -1$. Thus our formula reduces to

    $$\frac{1}{x} \approx 1 - (x-1) = 2 - x.$$

    This is the local linearization of $1/x$ near $x = 1$.

5. (a) Let $f(x) = (1+x)^k$. Then $f'(x) = k(1+x)^{k-1}$. Since

    $$f(x) \approx f(0) + f'(0)(x-0)$$

    is the tangent line approximation, and $f(0) = 1$, $f'(0) = k$, for small x we get

    $$f(x) \approx 1 + kx.$$

    (b) Since $\sqrt{1.1} = (1+0.1)^{1/2} \approx 1 + (1/2)0.1 = 1.05$ by the above method, this estimate is about right.

    (c) The real answer is less than 1.05. Since $(1.05)^2 = (1+0.05)^2 = 1+2(1)(0.05) + (0.05)^2 = 1.1 + (0.05)^2 > 1.1$, we have $(1.05)^2 > 1.1$ Therefore

    $$\sqrt{1.1} < 1.05.$$

    Graphically, we can see this because the graph of $\sqrt{1+x}$ is concave down, so it bends below its tangent line. (See Figure 4.8 on page 228.) Therefore the true value $(\sqrt{1.1})$ which is on the curve is below the approximate value (1.05) which is on the tangent line.

9. Note that $f(0) = g(0) = 0$ and $f'(0) = g'(0)$. Since $x = 0$ looks like a point of inflection for each curve, $f''(0) = g''(0) = 0$. Therefore, applying l'Hopital's rule successively gives us

    $$\lim_{x \to 0} \frac{f(x)}{g(x)} = \lim_{x \to 0} \frac{f'(x)}{g'(x)} = \lim_{x \to 0} \frac{f''(x)}{g''(x)} = \lim_{x \to 0} \frac{f'''(x)}{g'''(x)}.$$

    Now notice how the concavity of f changes: for $x < 0$, it is concave up, so $f''(x) > 0$, and for $x > 0$ it is concave down, so $f''(x) < 0$. Thus $f''(x)$ is a decreasing function at 0 and so $f'''(0)$ is negative. Similarly, for $x < 0$, we see g is concave down and for $x > 0$ it is concave up, so $g''(x)$ is increasing at 0 and so $g'''(0)$ is positive. Consequently,

    $$\lim_{x \to 0} \frac{f(x)}{g(x)} = \lim_{x \to 0} \frac{f'''(x)}{g'''(0)} < 0.$$

13.   The larger power dominates. Using l'Hopital's rule

$$\lim_{x\to\infty}\frac{x^5}{0.1x^7} = \lim_{x\to\infty}\frac{5x^4}{0.7x^6} = \lim_{x\to\infty}\frac{20x^3}{4.2x^5}$$

$$= \lim_{x\to\infty}\frac{60x^2}{21x^4} = \lim_{x\to\infty}\frac{120x}{84x^3} = \lim_{x\to\infty}\frac{120}{252x^2} = 0$$

so $0.1x^7$ dominates.

17.   Observe that both $f(4)$ and $g(4)$ are zero. Also, $f'(4) = 1.4$ and $g'(4) = -0.7$, so by l'Hopital's rule,

$$\lim_{x\to 4}\frac{f(x)}{g(x)} = \frac{f'(4)}{g'(4)} = \frac{1.4}{-0.7} = -2.$$

21.   (a)   Let $f(x) = 1/(1+x)$. Then $f'(x) = -1/(1+x)^2$ by the chain rule. So $f(0) = 1$, and $f'(0) = -1$. Therefore, for x near 0, $1/(1+x) \approx f(0) + f'(0)x = 1 - x$.

   (b)   We know that for small y, $1/(1+y) \approx 1 - y$. Let $y = x^2$; when x is small, so is $y = x^2$. Hence, for small x, $1/(1+x^2) \approx 1 - x^2$.

   (c)   Since the linearization of $1/(1+x^2)$ is the line $y = 1$, and this line has a slope of 0, the derivative of $1/(1+x^2)$ is zero at $x = 0$.

Solutions for Chapter 4 Review

1.   We wish to find the slope $m = dy/dx$. To do this, we can implicitly differentiate the given formula in terms of x:

$$x^2 + 3y^2 = 7$$
$$2x + 6y\frac{dy}{dx} = \frac{d}{dx}(7) = 0$$
$$\frac{dy}{dx} = \frac{-2x}{6y} = \frac{-x}{3y}.$$

Thus, at $(2, -1)$, $m = -(2)/3(-1) = 2/3$.

5.   When we zoom in on the origin, we find that two functions are not defined there. The other functions all look like straight lines through the origin. The only way we can tell them apart is their slope.

   The following functions all have slope 0 and are therefore indistinguishable:

$\sin x - \tan x$, $\frac{x^2}{x^2+1}$, $x - \sin x$, and $\frac{1-\cos x}{\cos x}$.

   These functions all have slope 1 at the origin, and are thus indistinguishable:

$\arcsin x$, $\frac{\sin x}{1+\sin x}$, $\arctan x$, $e^x - 1$, $\frac{x}{x+1}$, and $\frac{x}{x^2+1}$.

   Now, $\frac{\sin x}{x} - 1$ and $-x\ln x$ both are undefined at the origin, so they are distinguishable from the other functions. In addition, while $\frac{\sin x}{x} - 1$ has a slope that approaches zero near the origin, $-x\ln x$ becomes vertical near the origin, so they are distinguishable from each other.

   Finally, $x^{10} + \sqrt[10]{x}$ is the only function defined at the origin and with a vertical tangent there, so it is distinguishable from the others.

9.   (a)   If the distance $s(t) = 20e^{\frac{t}{2}}$, then the velocity, $v(t)$, is given by

$$v(t) = s'(t) = \left(20e^{\frac{t}{2}}\right)' = \left(\frac{1}{2}\right)\left(20e^{\frac{t}{2}}\right) = 10e^{\frac{t}{2}}.$$

   (b)   Observing the differentiation in (a), we note that

$$s'(t) = v(t) = \frac{1}{2}\left(20e^{\frac{t}{2}}\right) = \frac{1}{2}s(t).$$

Substituting $s(t)$ for $20e^{\frac{t}{2}}$, we obtain $s'(t) = \frac{1}{2}s(t)$.

13.   (a)   Differentiating, we see

$$v = \frac{dy}{dt} = -2\pi\omega y_0 \sin(2\pi\omega t)$$

$$a = \frac{dv}{dt} = -4\pi^2\omega^2 y_0 \cos(2\pi\omega t).$$

(b)   We have

$$y = y_0 \cos(2\pi\omega t)$$
$$v = -2\pi\omega y_0 \sin(2\pi\omega t)$$
$$a = -4\pi^2\omega^2 y_0 \cos(2\pi\omega t).$$

So

Amplitude of y is $|y_0|$,

Amplitude of v is $|2\pi\omega y_0| = 2\pi\omega|y_0|$,

Amplitude of a is $|4\pi^2\omega^2 y_0| = 4\pi^2\omega^2|y_0|$.

The amplitudes are different (provided $2\pi\omega \neq 1$). The periods of the three functions are all the same, namely $1/\omega$.

(c)   Looking at the answer to part (a), we see

$$\frac{d^2y}{dt^2} = a = -4\pi^2\omega^2 \left(y_0 \cos(2\pi\omega t)\right)$$

$$= -4\pi^2\omega^2 y.$$

So we see that

$$\frac{d^2y}{dt^2} + 4\pi^2\omega^2 y = 0.$$

17.   Using Pythagoras' theorem, we see that the distance x between the aircraft's current position and the point 2 miles directly above the ground station are related to s by the formula $x = (s^2 - 2^2)^{1/2}$. See Figure 4.4. The speed along the aircraft's constant altitude flight path is

$$\frac{dx}{dt} = \left(\frac{1}{2}\right)(s^2 - 4)^{-1/2}(2s)\left(\frac{ds}{dt}\right) = \frac{s}{x}\frac{ds}{dt}.$$

When $s = 4.6$ and $ds/dt = 210$,

$$\frac{dx}{dt} = \frac{4.6}{\sqrt{(4.6)^2 - 4}}210$$

$$= \frac{966}{\sqrt{21.16 - 4}}$$

$$= \frac{966}{4.14} \approx 233.2 \text{ miles/hour.}$$

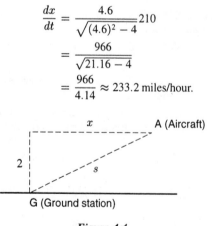

Figure 4.4

Solutions to Practice Problems on Differentiation

1. $f'(t) = 6t - 4$.

5. The power rule gives $f'(x) = 20x^3 - \dfrac{2}{x^3}$.

9. $y' = 2\left(\dfrac{x^2+2}{3}\right)\left(\dfrac{2x}{3}\right) = \dfrac{4}{9}x\left(x^2+2\right)$

13. $f(z) = \dfrac{z}{3} + \dfrac{1}{3}z^{-1} = \dfrac{1}{3}\left(z + z^{-1}\right)$, so $f'(z) = \dfrac{1}{3}\left(1 - z^{-2}\right) = \dfrac{1}{3}\left(\dfrac{z^2-1}{z^2}\right)$.

17. $j'(x) = \dfrac{ae^{ax}}{(e^{ax}+b)}$

21. $f'(x) = \cos(\cos x + \sin x)(\cos x - \sin x)$

25. $g'(t) = -4(3+\sqrt{t})^{-2}\left(\dfrac{1}{2}t^{-1/2}\right) = \dfrac{-2}{\sqrt{t}(3+\sqrt{t})^2}$

29. $q'(\theta) = \dfrac{1}{2}(4\theta^2 - \sin^2(2\theta))^{-1/2}(8\theta - 2\sin(2\theta)(2\cos(2\theta))) = \dfrac{4\theta - 2\sin(2\theta)\cos(2\theta)}{\sqrt{4\theta^2 - \sin^2(2\theta)}}$

33. Note that $g(x) = \arcsin(\sin \pi x) = \pi x$.
    Thus, $g'(x) = \pi$.

37. We can write $h(x) = \left(\dfrac{x^2+9}{x+3}\right)^{1/2}$, so

    $$h'(x) = \dfrac{1}{2}\left(\dfrac{x^2+9}{x+3}\right)^{-1/2}\left[\dfrac{2x(x+3) - (x^2+9)}{(x+3)^2}\right] = \dfrac{1}{2}\sqrt{\dfrac{x+3}{x^2+9}}\left[\dfrac{x^2+6x-9}{(x+3)^2}\right].$$

41. Using the product rule gives $v'(t) = 2te^{-ct} - ce^{-ct}t^2 = (2t - ct^2)e^{-ct}$.

45. Since $\ln\left[\left(\dfrac{1-\cos t}{1+\cos t}\right)^4\right] = 4\ln\left[\left(\dfrac{1-\cos t}{1+\cos t}\right)\right]$ we have

    $$a'(t) = 4\left(\dfrac{1+\cos t}{1-\cos t}\right)\left[\dfrac{\sin t(1+\cos t) + \sin t(1-\cos t)}{(1+\cos t)^2}\right]$$

    $$= \left[\dfrac{1+\cos t}{1-\cos t}\right]\left[\dfrac{8\sin t}{(1+\cos t)^2}\right]$$

    $$= \dfrac{8\sin t}{1-\cos^2 t}$$

    $$= \dfrac{8}{\sin t}.$$

49. $y' = (\ln \pi)\pi^{(x+2)}$.

53. $f'(x) = 2e^{2x}[x^2 + 5^x] + e^{2x}[2x + (\ln 5)5^x] = e^{2x}[2x^2 + 2x + (\ln 5 + 2)5^x]$.

57. $h'(z) = \dfrac{-8b^4 z}{(a+z^2)^5}$

61. $f'(x) = -\sin(\arctan 3x)\left(\dfrac{1}{1+(3x)^2}\right)(3) = \dfrac{-3\sin(\arctan 3x)}{1+9x^2}$.

65. $h'(r) = \dfrac{d}{dr}\left(\dfrac{r^2}{2r+1}\right) = \dfrac{(2r)(2r+1) - 2r^2}{(2r+1)^2} = \dfrac{2r(r+1)}{(2r+1)^2}$.

69. $g'(w) = \dfrac{d}{dw}\left(\dfrac{1}{2^w + e^w}\right) = -\dfrac{2^w \ln 2 + e^w}{(2^w + e^w)^2}$.

73. $r'(\theta) = \dfrac{d}{d\theta}\sin[(3\theta - \pi)^2] = \cos[(3\theta - \pi)^2] \cdot 2(3\theta - \pi) \cdot 3 = 6(3\theta - \pi)\cos[(3\theta - \pi)^2]$.

77. $w'(\theta) = \dfrac{1}{\sin^2 \theta} - \dfrac{2\theta \cos \theta}{\sin^3 \theta}$

81. Using the chain rule, we get:

$$m'(n) = \cos(e^n) \cdot (e^n)$$

85. $y' = 0$

89. $g'(x) = \dfrac{6x}{1 + \left(3x^2 + 1\right)^2} = \dfrac{6x}{9x^4 + 6x^2 + 2}$

93. $f'(\theta) = ke^{k\theta}$

97. Using the quotient rule gives

$$f'(x) = \frac{(-2x)(a^2 + x^2) - (2x)(a^2 - x^2)}{(a^2 + x^2)^2}$$
$$= \frac{-4a^2 x}{(a^2 + x^2)^2}.$$

101. Using the product rule gives

$$H'(t) = 2ate^{-ct} - c(at^2 + b)e^{-ct}$$
$$= (-cat^2 + 2at - bc)e^{-ct}.$$

105. $g'(u) = \dfrac{ae^{au}}{a^2 + b^2}$

109. $f'(x) = \dfrac{d}{dx}(2 - 4x - 3x^2)(6x^e - 3\pi) = (-4 - 6x)(6x^e - 3\pi) + (2 - 4x - 3x^2)(6ex^{e-1}).$

113.

$$h'(x) = \left(-\frac{1}{x^2} + \frac{2}{x^3}\right)\left(2x^3 + 4\right) + \left(\frac{1}{x} - \frac{1}{x^2}\right)\left(6x^2\right)$$
$$= -2x + 4 - \frac{4}{x^2} + \frac{8}{x^3} + 6x - 6$$
$$= 4x - 2 - 4x^{-2} + 8x^{-3}$$

117.

$$y + \frac{xdy}{dx} - 1 - \frac{3dy}{dx} = 0$$
$$(x - 3)\frac{dy}{dx} = 1 - y$$
$$\frac{dy}{dx} = \frac{1 - y}{x - 3}$$

121.

$$3x^2 + 3y^2\frac{dy}{dx} - 8xy - 4x^2\frac{dy}{dx} = 0$$
$$(3y^2 - 4x^2)\frac{dy}{dx} = 8xy - 3x^2$$
$$\frac{dy}{dx} = \frac{8xy - 3x^2}{3y^2 - 4x^2}$$

CHAPTER FIVE

Solutions for Section 5.1

1.

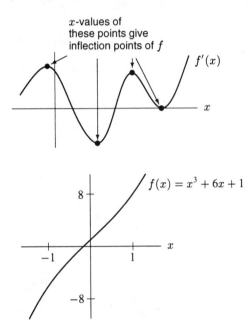

5. To find inflection points of the function f we must find points where f'' changes sign. However, because f'' is the derivative of f', any point where f'' changes sign will be a local maximum or minimum on the graph of f'.

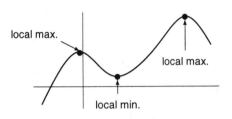

9.

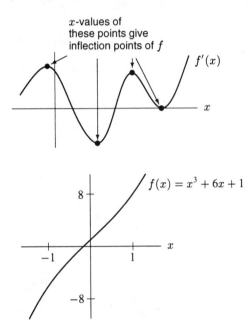

The graph of f in above appears to be increasing for all x, with no critical points. Since $f'(x) = 3x^2 + 6$ and $x^2 \geq 0$ for all x, we have $f'(x) > 0$ for all x. That explains why f is increasing for all x.

13.

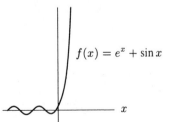

The graph of f above looks like $\sin x$ for $x < 0$ and e^x for $x > 0$. In particular, there are no waves for $x > 0$. We have $f'(x) = \cos x + e^x$, and so the critical points of f occur at those values of x for which $\cos x = -e^x$. Since $e^x > 1$ for all $x > 0$, we know immediately that there are no critical points at positive values of x. The specific locations of the critical points at $x < 0$ must be determined numerically; the first few are $x \approx -1.7, -4.7, -7.9$. For $x < 0$, the quantity e^x is small so that the graph looks like the graph of $\sin x$. For $x > 0$, we have $f'(x) > 0$ since $-1 \leq \cos x$ and $e^x > 1$. Thus, the graph is increasing for all $x > 0$ and there are no such waves.

17. Figure 5.1 contains the graph of $f(x) = x^2 + \cos x$.

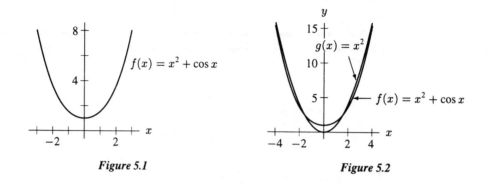

Figure 5.1   Figure 5.2

The graph looks like a parabola with no waves because $f''(x) = 2 - \cos x$, which is always positive. Thus, the graph of f is concave up everywhere; there are no waves. If you plot the graph of $f(x)$ together with the graph of $g(x) = x^2$, you see that the graph of f does wave back and forth across the graph of g, but never enough to change the concavity of f. See Figure 5.2.

21. Since f is differentiable everywhere, f' must be zero (not undefined) at any critical points; thus, $f'(3) = 0$. Since f has exactly one critical point, f' may change sign only at $x = 3$. Thus f is always increasing or always decreasing for $x < 3$ and for $x > 3$. Using the information in parts (a) through (d), we determine whether $x = 3$ is a local minimum, local maximum, or neither.

(a) $x = 3$ is a local (as well as a global) maximum because $f(x)$ is increasing when $x < 3$ and decreasing when $x > 3$.

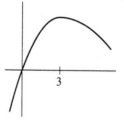

(b) $x = 3$ is a local (as well as a global) minimum because $f(x)$ heads to infinity to either side of $x = 3$.

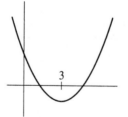

(c) $x = 3$ is neither a local minimum nor maximum, as $f(1) < f(2) < f(4) < f(5)$.

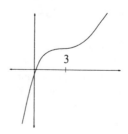

(d) $x = 3$ is a local (as well as a global) minimum because $f(x)$ is decreasing to the left of $x = 3$ and must increase to the right of $x = 3$, as $f(3) = 1$ and eventually $f(x)$ must become close to 3.

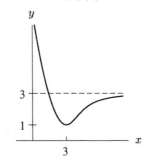

25.

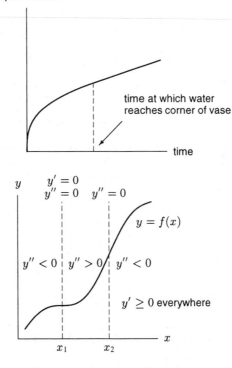

29.

33. (a) When a number grows larger, its reciprocal grows smaller. Therefore, since f is increasing near x_0, we know that g (its reciprocal) must be decreasing. Another argument can be made using derivatives. We know that (since f is increasing) $f'(x) > 0$ near x_0. We also know (by the chain rule) that $g'(x) = (f(x)^{-1})' = -\frac{f'(x)}{f(x)^2}$. Since both $f'(x)$ and $f(x)^2$ are positive, this means $g'(x)$ is negative, which in turn means $g(x)$ is decreasing near $x = x_0$.

(b) Since f has a local maximum near x_1, $f(x)$ increases as x nears x_1, and then $f(x)$ decreases as x exceeds x_1. Thus the reciprocal of f, g, decreases as x nears x_1 and then increases as x exceeds x_1. Thus g has a local minimum at $x = x_1$. To put it another way, since f has a local maximum at $x = x_1$, we know $f'(x_1) = 0$. Since $g'(x) = -\frac{f'(x)}{f(x)^2}$, $g'(x_1) = 0$. To the left of x_1, $f'(x_1)$ is positive, so $g'(x)$ is negative. To the right of x_1, $f'(x_1)$ is negative, so $g'(x)$ is positive. Therefore, g has a local minimum at x_1.

(c) Since f is concave down at x_2, we know $f''(x_2) < 0$. We also know (from above) that

$$g''(x_2) = \frac{2f'(x_2)^2}{f(x_2)^3} - \frac{f''(x_2)}{f(x_2)^2} = \frac{1}{f(x_2)^2}\left(\frac{2f'(x_2)^2}{f(x_2)} - f''(x_2)\right).$$

Since $\frac{1}{f(x_2)^2} > 0$, $2f'(x_2)^2 > 0$, and $f(x_2) > 0$ (as f is assumed to be everywhere positive), we see that $g''(x_2)$ is positive. Thus g is concave up at x_2.

Note that for the first two parts of the problem, we didn't need to require f to be positive (only non-zero). However, it was necessary here.

Solutions for Section 5.2

1. (a) Let $p(x) = x^3 - ax$, and suppose $a < 0$. Then $p'(x) = 3x^2 - a > 0$ for all x, so $p(x)$ is always increasing.

   (b) Now suppose $a > 0$. We have $p'(x) = 3x^2 - a = 0$ when $x^2 = a/3$, i.e., when $x = \sqrt{a/3}$ and $x = -\sqrt{a/3}$. We also have $p''(x) = 6x$; so $x = \sqrt{a/3}$ is a local minimum since $6\sqrt{a/3} > 0$, and $x = -\sqrt{a/3}$ is a local maximum since $-6\sqrt{a/3} < 0$.

   (c) Case 1: $a < 0$
   In this case, $p(x)$ is always increasing. We have $p''(x) = 6x > 0$ if $x > 0$, meaning the graph is concave up for $x > 0$. Furthermore, $6x < 0$ if $x < 0$, meaning the graph is concave down for $x < 0$. Thus, $x = 0$ is an inflection point.
   Case 2: $a > 0$
   We have

   $$p\left(\sqrt{\frac{a}{3}}\right) = \left(\sqrt{\frac{a}{3}}\right)^3 - a\sqrt{\frac{a}{3}} = \frac{a\sqrt{a}}{\sqrt{27}} - \frac{a\sqrt{a}}{\sqrt{3}} = -\frac{2a\sqrt{a}}{3\sqrt{3}} < 0,$$

   $$\text{and} \quad p\left(-\sqrt{\frac{a}{3}}\right) = -\frac{a\sqrt{a}}{\sqrt{27}} + \frac{a\sqrt{a}}{\sqrt{3}} = -p\left(\sqrt{\frac{a}{3}}\right) > 0.$$

   $$p'(x) = 3x^2 - a \begin{cases} = 0 & \text{if } |x| = \sqrt{\frac{a}{3}}; \\ > 0 & \text{if } |x| > \sqrt{\frac{a}{3}}; \\ < 0 & \text{if } |x| < \sqrt{\frac{a}{3}}. \end{cases}$$

   So p is increasing for $x < -\sqrt{a/3}$, decreasing for $-\sqrt{a/3} < x < \sqrt{a/3}$, and increasing for $x > \sqrt{a/3}$. Since $p''(x) = 6x$, the graph of $p(x)$ is concave down for values of x less than zero and concave up for values greater than zero. Graphs of $p(x)$ for $a < 0$ and $a > 0$ are found in Figures 5.3 and 5.4, respectively.

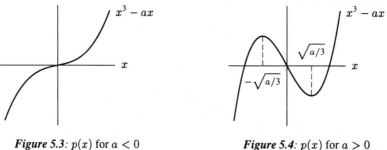

Figure 5.3: $p(x)$ for $a < 0$ **Figure 5.4**: $p(x)$ for $a > 0$

5. $T(t) =$ the temperature at time $t = a(1 - e^{-kt}) + b$.

   (a) Since at time $t = 0$ the yam is at $20°$C, we have

   $$T(0) = 20° = a\left(1 - e^0\right) + b = a(1 - 1) + b = b.$$

   Thus $b = 20°$C. Now, common sense tells us that after a period of time, the yam will heat up to about $200°$, or oven temperature. Thus the temperature T should approach $200°$ as the time t grows large:

   $$\lim_{t \to \infty} T(t) = 200°\text{C} = a(1 - 0) + b = a + b.$$

   Since $a + b = 200°$, and $b = 20°$C, this means $a = 180°$C.

   (b) Since we're talking about how quickly the yam is heating up, we need to look at the derivative, $T'(t) = ake^{-kt}$:

   $$T'(t) = (180)ke^{-kt}.$$

   We know $T'(0) = 2°$C/min, so

   $$2 = (180)ke^{-k(0)} = (180)(k).$$

   So $k = (2°\text{C/min})/180°\text{C} = \frac{1}{90}\text{min}^{-1}$.

9.

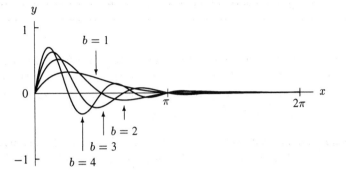

The larger the value of b, the narrower the humps and more humps per given region there are in the graph.

13. Let $f(x) = Ae^{-Bx^2}$. Since

$$f(x) = Ae^{-Bx^2} = Ae^{-\frac{(x-0)^2}{(1/B)}},$$

this is just the family of curves $y = e^{\frac{(x-a)^2}{b}}$ multiplied by a constant A. This family of curves is discussed in the text; here, $a = 0$, $b = \frac{1}{B}$. When $x = 0$, $y = Ae^0 = A$, so A determines the y-intercept. A also serves to flatten or stretch the graph of e^{-Bx^2} vertically. Since $f'(x) = -2ABxe^{-Bx^2}$, $f(x)$ has a critical point at $x = 0$. For $B > 0$, the graphs are bell-shaped curves centered at $x = 0$, and $f(0) = A$ is a global maximum.

To find the inflection points of f, we solve $f''(x) = 0$. Since $f'(x) = -2ABxe^{-Bx^2}$,

$$f''(x) = -2ABe^{-Bx^2} + 4AB^2x^2e^{-Bx^2}.$$

Since e^{-Bx^2} is always positive, $f''(x) = 0$ when

$$-2AB + 4AB^2x^2 = 0$$
$$x^2 = \frac{2AB}{4AB^2}$$
$$x = \pm\sqrt{\frac{1}{2B}}.$$

These are points of inflection, since the second derivative changes sign here. Thus for large values of B, the inflection points are close to $x = 0$, and for smaller values of B the inflection points are further from $x = 0$. Therefore B affects the width of the graph.

In the graphs in Figure 5.5, A is held constant, and variations in B are shown.

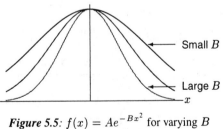

Figure 5.5: $f(x) = Ae^{-Bx^2}$ for varying B

17. (a) The larger the value of $|A|$, the steeper the graph (for the same x-value).
    (b) The graph is shifted horizontally by B. The shift is to the left for positive B, to the right for negative B. There is a vertical asymptote at $x = -B$. See Figure 5.6.

(c)

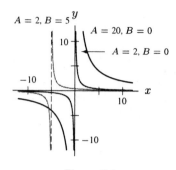

Figure 5.6

Solutions for Section 5.3

1.

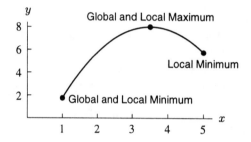

5. (a)

$$f(x) = \sin^2 x - \cos x \quad \text{for } 0 \le x \le \pi$$
$$f'(x) = 2\sin x \cos x + \sin x = (\sin x)(2\cos x + 1)$$

$f'(x) = 0$ when $\sin x = 0$ or when $2\cos x + 1 = 0$. Now, $\sin x = 0$ when $x = 0$ or when $x = \pi$. On the other hand, $2\cos x + 1 = 0$ when $\cos x = -1/2$, which happens when $x = 2\pi/3$. So the critical points are $x = 0$, $x = 2\pi/3$, and $x = \pi$.

Note that $\sin x > 0$ for $0 < x < \pi$. Also, $2\cos x + 1 < 0$ if $2\pi/3 < x \le \pi$ and $2\cos x + 1 > 0$ if $0 < x < 2\pi/3$. Therefore,

$$f'(x) < 0 \quad \text{for} \quad \frac{2\pi}{3} < x < \pi$$
$$f'(x) > 0 \quad \text{for} \quad 0 < x < \frac{2\pi}{3}.$$

Thus f has a local maximum at $x = 2\pi/3$ and local minima at $x = 0$ and $x = \pi$.

(b) We have

$$f(0) = [\sin(0)]^2 - \cos(0) = -1$$
$$f\left(\frac{2\pi}{3}\right) = \left[\sin\left(\frac{2\pi}{3}\right)\right]^2 - \cos\frac{2\pi}{3} = 1.25$$
$$f(\pi) = [\sin(\pi)]^2 - \cos(\pi) = 1.$$

Thus the global maximum is at $x = 2\pi/3$, and the global minimum is at $x = 0$.

9. We want to maximize the height, y, of the grapefruit above the ground, as shown in the figure below. Using the derivative we can find exactly when the grapefruit is at the highest point. We can think of this in two ways. By common sense, at the peak of the grapefruit's flight, the velocity, dy/dt, must be zero. Alternately, we are looking for a global maximum of y, so we look for critical points where $dy/dt = 0$. We have

$$\frac{dy}{dt} = -32t + 50 = 0 \quad \text{and so} \quad t = \frac{-50}{-32} \approx 1.56 \text{ sec.}$$

Thus, we have the time at which the height is a maximum; the maximum value of y is then

$$y \approx -16(1.56)^2 + 50(1.56) + 5 = 44.1 \text{ feet}.$$

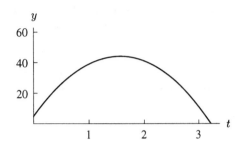

13.

$$\frac{dE}{d\theta} = \frac{(\mu + \theta)(1 - 2\mu\theta) - (\theta - \mu\theta^2)}{(\mu + \theta)^2} = \frac{\mu(1 - 2\mu\theta - \theta^2)}{(\mu + \theta)^2}.$$

Now $dE/d\theta = 0$ when $\theta = -\mu \pm \sqrt{1 + \mu^2}$. Since $\theta > 0$, the only possible critical point is when $\theta = -\mu + \sqrt{\mu^2 + 1}$. Differentiating again gives $E'' < 0$ at this point and so it is a local maximum. Since $E(\theta)$ is continuous for $\theta > 0$ and $E(\theta)$ has only one critical point, the local maximum is the global maximum.

17. (a) We want to find where $x > 2\ln x$, which is the same as solving $x - 2\ln x > 0$. Let $f(x) = x - 2\ln x$. Then $f'(x) = 1 - \frac{2}{x}$, which implies that $x = 2$ is the only critical point of f. Since $f'(x) < 0$ for $x < 2$ and $f'(x) > 0$ for $x > 2$, by the first derivative test we see that f has a local and global minimum at $x = 2$. Since $f(2) = 2 - 2\ln 2 \approx 0.61$, then for all $x > 0$, $f(x) \geq f(2) > 0$. Thus $f(x)$ is always positive, which means $x > 2\ln x$ for any $x > 0$.

   (b) We've shown that $x > 2\ln x = \ln(x^2)$ for all $x > 0$. Since e^x is an increasing function, $e^x > e^{\ln x^2} = x^2$, so $e^x > x^2$ for all $x > 0$.

   (c) Let $f(x) = x - 3\ln x$. Then $f'(x) = 1 - \frac{3}{x} = 0$ at $x = 3$. By the first derivative test, f has a local minimum at $x = 3$. But, $f(3) \approx -0.295$, which is less than zero. Thus $3\ln x > x$ at $x = 3$. So, x is not less than $3\ln x$ for all $x > 0$.

   (One could also see this by plugging in $x = e$: since $3\ln e = 3$, $x < 3\ln x$ when $x = e$.)

21. Let $y = x^3 - 4x^2 + 4x$. To locate the critical points, we solve $y' = 0$. Since $y' = 3x^2 - 8x + 4 = (3x - 2)(x - 2)$, the critical points are $x = 2/3$ and $x = 2$. To find the global minimum and maximum on $0 \leq x \leq 4$, we check the critical points and the endpoints: $y(0) = 0$; $y(2/3) = 32/27$; $y(2) = 0$; $y(4) = 16$. Thus, the global minimum is at $x = 0$ and $x = 2$, the global maximum is at $x = 4$, and $0 \leq y \leq 16$.

25. (a) Figure 5.7 contains the graph of total drag, plotted on the same coordinate system with induced and parasite drag. It was drawn by adding the vertical coordinates of Induced and Parasite drag.

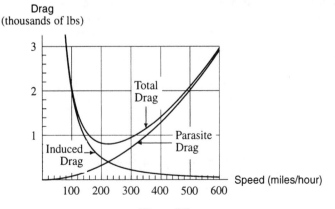

Figure 5.7

(b) Airspeeds of approximately 160 mph and 320 mph each result in a total drag of 1000 pounds. Since two distinct airspeeds are associated with a single total drag value, the total drag function does not have an inverse. The parasite and induced drag functions do have inverses, because they are strictly increasing and strictly decreasing functions, respectively.

(c) To conserve fuel, fly the at the airspeed which minimizes total drag. This is the airspeed corresponding to the lowest point on the total drag curve in part (a): that is, approximately 220 mph.

Solutions for Section 5.4

1.

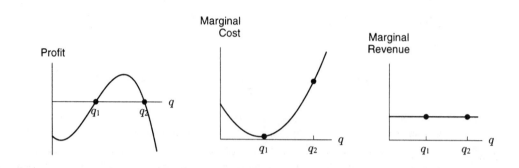

5. (a) The fixed cost is 0 because $C(0) = 0$.

(b) Profit, $\pi(q)$, is equal to money from sales, $7q$, minus total cost to produce those items, $C(q)$.

$$\pi = 7q - 0.01q^3 + 0.6q^2 - 13q$$
$$\pi' = -0.03q^2 + 1.2q - 6 = 0$$
$$q = \frac{-1.2 \pm \sqrt{(1.2)^2 - 4(0.03)(6)}}{-0.06} \approx 5.9 \text{ or, } 34.1.$$

Now $\pi'' = -0.06q + 1.2$, so $\pi''(5.9) > 0$ and $\pi''(34.1) < 0$. This means $q = 5.9$ is a local min and $q = 34.1$ a local max. We now evaluate the endpoint, $\pi(0) = 0$, and the points nearest $q = 34.1$ with integer q-values:

$$\pi(35) = 7(35) - 0.01(35)^3 + 0.6(35)^2 - 13(35) = 245 - 148.75 = 96.25,$$

$$\pi(34) = 7(34) - 0.01(34)^3 + 0.6(34)^2 - 13(34) = 238 - 141.44 = 96.56.$$

So the (global) maximum profit is $\pi(34) = 96.56$. The money from sales is $238, the cost to produce the items is $141.44, resulting in a profit of $96.56.

(c) The money from sales is equal to price×quantity sold. If the price is raised from $7 by x to $(7 + x)$, the result is a reduction in sales from 34 items to $(34 - 2x)$ items. So the result of raising the price by x is to change the money from sales from $(7)(34)$ to $(7 + x)(34 - 2x)$ dollars. If the production level is fixed at 34, then the production costs are fixed at $141.44, as found in part (b), and the profit is given by:

$$\pi(x) = (7 + x)(34 - 2x) - 141.44$$

This expression gives the profit as a function of change in price x, rather than as a function of quantity as in part (b). We set the derivative of π with respect to x equal to zero to find the change in price that maximizes the profit:

$$\frac{d\pi}{dx} = (1)(34 - 2x) + (7 + x)(-2) = 20 - 4x = 0$$

So $x = 5$, and this must give a maximum for $\pi(x)$ since the graph of π is a parabola which opens downwards. The profit when the price is $12 (= 7 + x = 7 + 5)$ is thus $\pi(5) = (7 + 5)(34 - 2(5)) - 141.44 = $146.56. This is indeed higher than the profit when the price is $7, so the smart thing to do is to raise the price by $5.

9. (a) The value of MC is the slope of the tangent to the curve at q_0. See Figure 5.8.
    (b) The line from the curve to the origin joins $(0, 0)$ and $(q_0, C(q_0))$, so its slope is $C(q_0)/q_0 = a(q_0)$.
    (c) Figure 5.9 shows that the line whose slope is the minimum $a(q)$ is tangent to the curve $C(q)$. This line, therefore, also has slope MC, so $a(q) = MC$ at the q making $a(q)$ minimum.

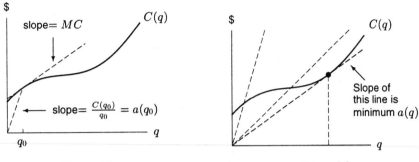

Figure 5.8   **Figure 5.9**

Solutions for Section 5.5

1. We take the derivative, set it equal to 0, and solve for x:

$$\frac{dt}{dx} = \frac{1}{6} - \frac{1}{4} \cdot \frac{1}{2} \left((2000 - x)^2 + 600^2 \right)^{-1/2} \cdot 2(2000 - x) = 0$$

$$(2000 - x) = \frac{2}{3} \left((2000 - x)^2 + 600^2 \right)^{1/2}$$

$$(2000 - x)^2 = \frac{4}{9} \left((2000 - x)^2 + 600^2 \right)$$

$$\frac{5}{9}(2000 - x)^2 = \frac{4}{9} \cdot 600^2$$

$$2000 - x = \sqrt{\frac{4}{5} \cdot 600^2} = \frac{1200}{\sqrt{5}}$$

$$x = 2000 - \frac{1200}{\sqrt{5}} \text{ feet.}$$

Note that $2000 - 1200/\sqrt{5} \approx 1463$ feet, as given in the example.

5. (a) Suppose the height of the box is h. The box has six sides, four with area xh and two, the top and bottom, with area x^2. Thus,

$$4xh + 2x^2 = A.$$

So

$$h = \frac{A - 2x^2}{4x}.$$

Then, the volume, V, is given by

$$V = x^2 h = x^2 \left(\frac{A - 2x^2}{4x} \right) = \frac{x}{4} \left(A - 2x^2 \right)$$

$$= \frac{A}{4} x - \frac{1}{2} x^3.$$

    (b) The graph is shown in Figure 5.10. We are assuming A is a positive constant. Also, we have drawn the whole graph, but we should only consider $V > 0$, $x > 0$ as V and x are lengths.

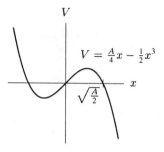

Figure 5.10

(c)   To find the maximum, we differentiate, regarding A as a constant:

$$\frac{dV}{dx} = \frac{A}{4} - \frac{3}{2}x^2.$$

So $dV/dx = 0$ if

$$\frac{A}{4} - \frac{3}{2}x^2 = 0$$

$$x = \pm\sqrt{\frac{A}{6}}.$$

For a real box, we must use $x = \sqrt{A/6}$. The graph makes it clear that this value of x gives the maximum. Evaluating at $x = \sqrt{A/6}$, we get

$$V = \frac{A}{4}\sqrt{\frac{A}{6}} - \frac{1}{2}\left(\sqrt{\frac{A}{6}}\right)^3 = \frac{A}{4}\sqrt{\frac{A}{6}} - \frac{1}{2} \cdot \frac{A}{6}\sqrt{\frac{A}{6}} = \left(\frac{A}{6}\right)^{3/2}.$$

9.   If the illumination is represented by I, then we know that

$$I = \frac{k\cos\theta}{r^2}.$$

See Figure 5.11.

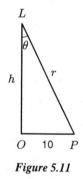

Figure 5.11

Since $r^2 = h^2 + 10^2$ and $\cos\theta = h/r = h/\sqrt{h^2 + 10^2}$, we have

$$I = \frac{kh}{(h^2 + 10^2)^{3/2}}.$$

To find the height at which I is maximized, we differentiate

$$\frac{dI}{dh} = \frac{k}{\left(h^2 + 10^2\right)^{3/2}} - \frac{3kh(2h)}{2\left(h^2 + 10^2\right)^{5/2}} = \frac{k(h^2 + 10^2) - 3kh^2}{\left(h^2 + 10^2\right)^{5/2}} = \frac{k\left(10^2 - 2h^2\right)}{\left(h^2 + 10^2\right)^{5/2}}.$$

Setting $dI/dh = 0$ gives

$$10^2 - 2h^2 = 0$$
$$h = \sqrt{50} \text{ meters.}$$

Since $dI/dh > 0$ for $0 \leq h < \sqrt{50}$ and $dI/dh < 0$ for $h > \sqrt{50}$, we know that I is a maximum when $h = \sqrt{50}$ meters.

13. Let the sides of the rectangle have lengths a and b. We shall look for the minimum of the square s of the length of either diagonal, i.e. $s = a^2 + b^2$. The area is $A = ab$, so $b = A/a$. This gives

$$s(a) = a^2 + \frac{A^2}{a^2}.$$

To find the minimum squared length we need to find the critical points of s. Differentiating s with respect to a gives

$$\frac{ds}{da} = 2a + (-2)A^2 a^{-3} = 2a\left(1 - \frac{A^2}{a^4}\right)$$

The derivative $ds/da = 0$ when $a = \sqrt{A}$, that is when $a = b$ and so the rectangle is a square. Because $\dfrac{d^2 s}{da^2} = 2\left(1 + \frac{3A^2}{a^4}\right) > 0$, this is a minimum.

17. (a) If, following the hint, we set $f(x) = (a+x)/2 - \sqrt{ax}$, then $f(x)$ represents the difference between the arithmetic and geometric means for some fixed a and any $x > 0$. We can find where this difference is minimized by solving $f'(x) = 0$. Since $f'(x) = \frac{1}{2} - \frac{1}{2}\sqrt{a}x^{-1/2}$, if $f'(x) = 0$ then $\frac{1}{2}\sqrt{a}x^{-1/2} = \frac{1}{2}$, or $x = a$. Since $f''(x) = \frac{1}{4}\sqrt{a}x^{-3/2}$ is positive for all positive x, by the second derivative test $f(x)$ has a minimum at $x = a$, and $f(a) = 0$. Thus $f(x) = (a+x)/2 - \sqrt{ax} \geq 0$ for all $x > 0$, which means $(a+x)/2 \geq \sqrt{ax}$. This means that the arithmetic mean is greater than the geometric mean unless $a = x$, in which case the two means are equal.

   Alternatively, and without using calculus, we obtain

$$\frac{a+b}{2} - \sqrt{ab} = \frac{a - 2\sqrt{ab} + b}{2}$$
$$= \frac{(\sqrt{a} - \sqrt{b})^2}{2} \geq 0,$$

   and again we have $(a+b)/2 \geq \sqrt{ab}$.

   (b) Following the hint, set $f(x) = \frac{a+b+x}{3} - \sqrt[3]{abx}$. Then $f(x)$ represents the difference between the arithmetic and geometric means for some fixed a, b and any $x > 0$. We can find where this difference is minimized by solving $f'(x) = 0$. Since $f'(x) = \frac{1}{3} - \frac{1}{3}\sqrt[3]{ab}x^{-2/3}$, $f'(x) = 0$ implies that $\frac{1}{3}\sqrt[3]{ab}x^{-2/3} = \frac{1}{3}$, or $x = \sqrt{ab}$. Since $f''(x) = \frac{2}{9}\sqrt[3]{ab}x^{-5/3}$ is positive for all positive x, by the second derivative test $f(x)$ has a minimum at $x = \sqrt{ab}$. But

$$f(\sqrt{ab}) = \frac{a+b+\sqrt{ab}}{3} - \sqrt[3]{ab\sqrt{ab}} = \frac{a+b+\sqrt{ab}}{3} - \sqrt{ab} = \frac{a+b-2\sqrt{ab}}{3}.$$

   By the first part of this problem, we know that $\frac{a+b}{2} - \sqrt{ab} \geq 0$, which implies that $a + b - 2\sqrt{ab} \geq 0$. Thus $f(\sqrt{ab}) = \frac{a+b-2\sqrt{ab}}{3} \geq 0$. Since f has a maximum at $x = \sqrt{ab}$, $f(x)$ is always nonnegative. Thus $f(x) = \frac{a+b+x}{3} - \sqrt[3]{abx} \geq 0$, so $\frac{a+b+c}{3} \geq \sqrt[3]{abc}$. Note that equality holds only when $a = b = c$. (Part (b) may also be done without calculus, but it's harder than (a).)

21. (a) Since $RB' = x$ and $A'R = c - x$, we have

$$AR = \sqrt{a^2 + (c-x)^2} \quad \text{and} \quad RB = \sqrt{b^2 + x^2}.$$

   See Figure 5.12.

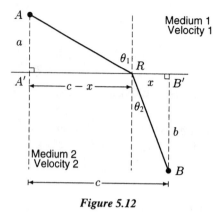

Figure 5.12

The time traveled, T, is given by

$$T = \text{Time } AR + \text{Time } RB = \frac{\text{Distance } AR}{v_1} + \frac{\text{Distance } RB}{v_2}$$

$$= \frac{\sqrt{a^2 + (c-x)^2}}{v_1} + \frac{\sqrt{b^2 + x^2}}{v_2}.$$

(b) Let us calculate dT/dx:

$$\frac{dT}{dx} = \frac{-2(c-x)}{2v_1\sqrt{a^2 + (c-x)^2}} + \frac{2x}{2v_2\sqrt{b^2 + x^2}}.$$

At the minimum $dT/dx = 0$, so

$$\frac{c-x}{v_1\sqrt{a^2 + (c-x)^2}} = \frac{x}{v_2\sqrt{b^2 + x^2}}.$$

But we have

$$\sin\theta_1 = \frac{c-x}{\sqrt{a^2 + (c-x)^2}} \quad \text{and} \quad \sin\theta_2 = \frac{x}{\sqrt{b^2 + x^2}}.$$

Therefore, setting $dT/dx = 0$ tells us that

$$\frac{\sin\theta_1}{v_1} = \frac{\sin\theta_2}{v_2}$$

which gives

$$\frac{\sin\theta_1}{\sin\theta_2} = \frac{v_1}{v_2}.$$

Solutions for Section 5.6

1. Substitute $x = 0$ into the formula for $\sinh x$. This yields

$$\sinh 0 = \frac{e^0 - e^{-0}}{2} = \frac{1-1}{2} = 0.$$

5. First, we observe that

$$\cosh(2x) = \frac{e^{2x} + e^{-2x}}{2}.$$

Now let's use the fact that $e^x \cdot e^{-x} = 1$ to calculate

$$\cosh^2 x = \left(\frac{e^x + e^{-x}}{2} \right)^2$$

$$= \frac{(e^x)^2 + 2e^x \cdot e^{-x} + (e^{-x})^2}{4}$$

$$= \frac{e^{2x} + 2 + e^{-2x}}{4}.$$

Similarly, we have

$$\sinh^2 x = \left(\frac{e^x - e^{-x}}{2} \right)^2$$

$$= \frac{(e^x)^2 - 2e^x \cdot e^{-x} + (e^{-x})^2}{4}$$

$$= \frac{e^{2x} - 2 + e^{-2x}}{4}.$$

Thus, to obtain $\cosh(2x)$, we need to add (rather than subtract) $\cosh^2 x$ and $\sinh^2 x$, giving

$$\cosh^2 x + \sinh^2 x = \frac{e^{2x} + 2 + e^{-2x} + e^{2x} - 2 + e^{-2x}}{4}$$

$$= \frac{2e^{2x} + 2e^{-2x}}{4}$$

$$= \frac{e^{2x} + e^{-2x}}{2}$$

$$= \cosh(2x).$$

Thus, we see that the identity relating $\cosh(2x)$ to $\cosh x$ and $\sinh x$ is

$$\cosh(2x) = \cosh^2 x + \sinh^2 x.$$

9. Using the chain rule twice, $\dfrac{d}{dt}\left(\cosh(e^{t^2}) \right) = \sinh(e^{t^2}) \cdot e^{t^2} \cdot 2t = 2te^{t^2} \sinh(e^{t^2})$.

13. (a) The graph below looks like the graph of $y = \cosh x$, with the minimum at about $(0.5, 6.3)$.

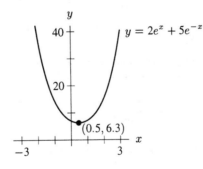

(b) We want to write

$$y = 2e^x + 5e^{-x} = A\cosh(x - c) = \frac{A}{2}e^{x-c} + \frac{A}{2}e^{-(x-c)}$$

$$= \frac{A}{2}e^x e^{-c} + \frac{A}{2}e^{-x}e^{c}$$

$$= \left(\frac{Ae^{-c}}{2} \right)e^x + \left(\frac{Ae^{c}}{2} \right)e^{-x}.$$

Thus, we need to choose A and c so that

$$\frac{Ae^{-c}}{2} = 2 \quad \text{and} \quad \frac{Ae^{c}}{2} = 5.$$

Dividing gives

$$\frac{Ae^{c}}{Ae^{-c}} = \frac{5}{2}$$
$$e^{2c} = 2.5$$
$$c = \frac{1}{2}\ln 2.5 \approx 0.458.$$

Solving for A gives

$$A = \frac{4}{e^{-c}} = 4e^{c} \approx 6.325.$$

Thus,

$$y = 6.325\cosh(x - 0.458).$$

Rewriting the function in this way shows that the graph in part (a) is the graph of $\cosh x$ shifted to the right by 0.458 and stretched vertically by a factor of 6.325.

17.

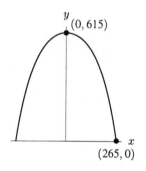

We know $x = 0$ and $y = 615$ at the top of the arch, so

$$615 = b - a\cosh(0/a) = b - a.$$

This means $b = a + 615$. We also know that $x = 265$ and $y = 0$ where the arch hits the ground, so

$$0 = b - a\cosh(265/a) = a + 615 - a\cosh(265/a).$$

We can solve this equation numerically on a calculator and get $a \approx 100$, which means $b \approx 715$. This results in the equation

$$y \approx 715 - 100\cosh\left(\frac{x}{100}\right).$$

Solutions for Chapter 5 Review

1.

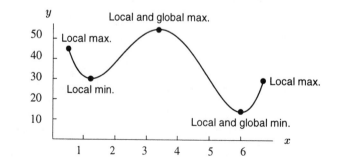

5. (a) $(-\infty, 0)$ decreasing, $(0, 4)$ increasing, $(4, \infty)$ decreasing.

  (b) local minimum at $f(0)$, local maximum at $f(4)$.

9. (a) First we find f' and f'':

$$f'(x) = -e^{-x}\sin x + e^{-x}\cos x$$
$$f''(x) = e^{-x}\sin x - e^{-x}\cos x$$
$$-e^{-x}\cos x - e^{-x}\sin x$$
$$= -2e^{-x}\cos x$$

  (b) The critical points are $x = \pi/4, 5\pi/4$, since $f'(x) = 0$ here.

  (c) The inflection points are $x = \pi/2, 3\pi/2$, since f'' changes sign at these points.

  (d) At the endpoints, $f(0) = 0$, $f(2\pi) = 0$. So we have $f(\pi/4) = (e^{-\pi/4})(\sqrt{2}/2)$ as the global maximum; $f(5\pi/4) = -e^{-5\pi/4}(\sqrt{2}/2)$ as the global minimum.

  (e)

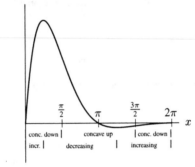

13. As $x \to -\infty$, $e^{-x} \to \infty$, so $xe^{-x} \to -\infty$. Thus $\lim_{x \to -\infty} xe^{-x} = -\infty$.

  As $x \to \infty$, $\frac{x}{e^x} \to 0$, since e^x grows much more quickly than x. Thus $\lim_{x \to \infty} xe^{-x} = 0$.

  Using the product rule,

$$f'(x) = e^{-x} - xe^{-x} = (1 - x)e^{-x},$$

which is zero when $x = 1$, negative when $x > 1$, and positive when $x < 1$. Thus $f(1) = 1/e^1 = 1/e$ is a local maximum.

  Again, using the product rule,

$$f''(x) = -e^{-x} - e^{-x} + xe^{-x}$$
$$= xe^{-x} - 2e^{-x}$$
$$= (x - 2)e^{-x},$$

which is zero when $x = 2$, positive when $x > 2$, and negative when $x < 2$, giving an inflection point at $(2, \frac{2}{e^2})$. With the above, we have the following diagram:

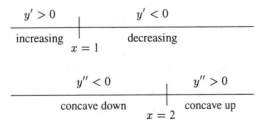

The graph of f is shown in Figure 5.13.

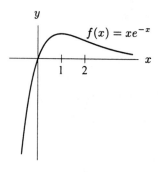

Figure 5.13

and $f(x)$ has one global maximum at $1/e$ and no local or global minima.

17. $\lim\limits_{x \to +\infty} f(x) = +\infty$, and $\lim\limits_{x \to 0^+} f(x) = +\infty$.

Hence, $x = 0$ is a vertical asymptote.

$f'(x) = 1 - \dfrac{2}{x} = \dfrac{x - 2}{x}$, so $x = 2$ is the only critical point.

$f''(x) = \dfrac{2}{x^2}$, which can never be zero. So there are no inflection points.

x		2	
f'	$-$	0	$+$
f''	$+$	$+$	$+$
f	$\searsmile$		$\nearsmile$

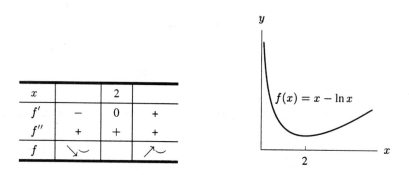

Thus, $f(2)$ is a local and global minimum.

21.   The critical points of f occur where f' is zero. These two points are indicated in the figure below.

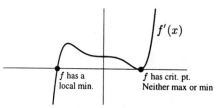

Note that the point labeled as a local minimum of f is not a critical point of f'.

25.   Volume: $V = x^2 y$,

Surface: $S = x^2 + 4xy = x^2 + 4xV/x^2 = x^2 + 4V/x$.

To find the dimensions which minimize the area, find x such that $dS/dx = 0$.

$$\frac{dS}{dx} = 2x - \frac{4V}{x^2} = 0,$$

so

$$x^3 = 2V,$$

and solving for x gives $x = \sqrt[3]{2V}$. To see that this gives a minimum, note that for small x, $S \approx 4V/x$ is decreasing. For large x, $S \approx x^2$ is increasing. Since there is only one critical point, this must give a global minimum. Using x to find y gives $y = V/x^2 = V/(2V)^{2/3} = \sqrt[3]{V/4}$.

29. (a) We have $g'(t) = \frac{t(1/t) - \ln t}{t^2} = \frac{1 - \ln t}{t^2}$, which is zero if $t = e$, negative if $t > e$, and positive if $t < e$, since $\ln t$ is increasing. Thus $g(e) = \frac{1}{e}$ is a global maximum for g. Since $t = e$ was the only point at which $g'(t) = 0$, there is no minimum.

(b) Now $\ln t/t$ is increasing for $0 < t < e$, $\ln 1/1 = 0$, and $\ln 5/5 \approx 0.322 < \ln(e)/e$. Thus, for $1 < t < e$, $\ln t/t$ increases from 0 to above $\ln 5/5$, so there must be a t between 1 and e such that $\ln t/t = \ln 5/5$. For $t > e$, there is only one solution to $\ln t/t = \ln 5/5$, namely $t = 5$, since $\ln t/t$ is decreasing for $t > e$. For $0 < t < 1$, $\ln t/t$ is negative and so cannot equal $\ln 5/5$. Thus $\ln x/x = \ln t/t$ has exactly two solutions.

(c) The graph of $\ln t/t$ intersects the horizontal line $y = \ln 5/5$, at $x = 5$ and $x \approx 1.75$.

33. (a) The concavity changes at t_1 and t_3, as shown below.

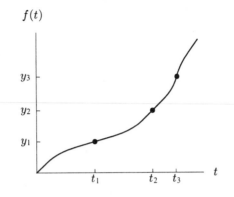

$f(t)$

(b) $f(t)$ grows most quickly where the vase is skinniest (at y_3) and most slowly where the vase is widest (at y_1). The diameter of the widest part of the vase looks to be about 4 times as large as the diameter at the skinniest part. Since the area of a cross section is given by πr^2, where r is the radius, the ratio between areas of cross sections at these two places is about 4^2, so the growth rates are in a ratio of about 1 to 16 (the wide part being 16 times slower).

Solutions to Problems on the Theorems about Continuous and Differentiable Functions ▬

1. Let $f(x) = \sin x$ and $g(x) = x$. Then $f(0) = 0$ and $g(0) = 0$. Also $f'(x) = \cos x$ and $g'(x) = 1$, so for all $x \geq 0$ we have $f'(x) \leq g'(x)$. So the graphs of f and g both go through the origin and the graph of f climbs slower than the graph of g. Thus the graph of f is below the graph of g for $x \geq 0$ by the Racetrack Principle. In other words, $\sin x \leq x$ for $x \geq 0$.

5. If f is continuous then $-f$ is continuous also. So $-f$ has a global maximum at some point $x = c$. Thus $-f(x) \leq -f(c)$ for all x in $[a, b]$. Hence $f(x) \geq f(c)$ for all x in $[a, b]$. So f has a global minimum at $x = c$.

9. Let $h(x) = f(x) - g(x)$. Then $h'(x) = f'(x) - g'(x) = 0$ for all x in (a, b). Hence, by the Constant Function Theorem, there is a constant C such that $h(x) = C$ on (a, b). Thus $f(x) = g(x) + C$.

13. (a) Since $f''(x) \geq 0$, $f'(x)$ is nondecreasing on (a, b). Thus $f'(c) \leq f'(x)$ for $c \leq x < b$ and $f'(x) \leq f'(c)$ for $a < x \leq c$.

(b) Let $g(x) = f(c) + f'(c)(x - c)$ and $h(x) = f(x)$. Then $g(c) = f(c) = h(c)$, and $g'(x) = f'(c)$ and $h'(x) = f'(x)$. If $c \leq x < b$, then $g'(x) \leq h'(x)$, and if $a < x \leq c$, then $g'(x) \geq h'(x)$, by (a). By the Racetrack Principle, $g(x) \leq h'(x)$ for $c \leq x < b$ and for $a < x \leq c$, as we wanted.

17. (a) We have $y < y + 1$. Since $y \geq 0$, we have $y + 1 > 0$, so dividing both sides by $y + 1$ we get $y/(y + 1) < 1$.

(b) It follows from the theorem on properties of continuous functions on page 132 that g is continuous. It follows from part (a) that g is bounded above by 1.

(c) Let $g(x) = x/(x + 1)$. Then,

$$g'(x) = \frac{1}{(x + 1)^2} > 0 \quad \text{for all } x \geq 0.$$

As a result, $g(y_1) \leq g(y_2)$ implies $y_1 \leq y_2$.

(d) Let c be the point where g has a global maximum. Then $g(x) \leq g(c)$ for all x in $[a, b]$, hence $f(x)/(1 + f(x)) \leq f(c)/(1 + f(c))$, and hence, by part (c), $f(x) \leq f(c)$ for all x in $[a, b]$. Hence f has a global maximum on $[a, b]$ at $x = c$.

(e) If f is continuous, then so is $|f|$. By what we have shown, $|f|$ has a global maximum U on $[a, b]$. But if $|f(x)| \leq U$, then $f(x) \leq |f(x)| \leq U$ also, so f is also bounded above.

CHAPTER SIX

Solutions for Section 6.1

1.

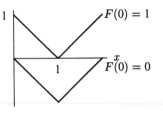

5. (a) Critical points of $F(x)$ are the zeros of f: $x = 1$ and $x = 3$.
   (b) $F(x)$ has a local minimum at $x = 1$ and a local maximum at $x = 3$.
   (c)

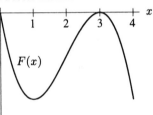

Notice that the graph could also be above or below the x-axis at $x = 3$.

9.

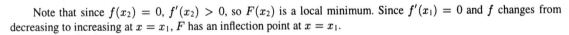

Note that since $f(x_2) = 0$, $f'(x_2) > 0$, so $F(x_2)$ is a local minimum. Since $f'(x_1) = 0$ and f changes from decreasing to increasing at $x = x_1$, F has an inflection point at $x = x_1$.

13. Looking at the graph of g' below, we see that the critical points of g occur when $x = 15$ and $x = 40$, since $g'(x) = 0$ at these values. Inflection points of g occur when $x = 10$ and $x = 20$, because $g'(x)$ has a local maximum or minimum at these values. Knowing these four key points, we sketch the graph of $g(x)$ as follows.

We start at $x = 0$, where $g(0) = 50$. Since g' is negative on the interval $[0, 10]$, the value of $g(x)$ is decreasing there. At $x = 10$ we have

$$g(10) = g(0) + \int_0^{10} g'(x)\, dx$$

$$= 50 - (\text{area of shaded trapezoid } T_1)$$

$$= 50 - \left(\frac{10 + 20}{2} \cdot 10\right) = -100.$$

Similarly,

$$g(15) = g(10) + \int_{10}^{15} g'(x)\, dx$$
$$= -100 - (\text{area of triangle } T_2)$$
$$= -100 - \frac{1}{2}(5)(20) = -150.$$

Continuing,

$$g(20) = g(15) + \int_{15}^{20} g'(x)\, dx = -150 + \frac{1}{2}(5)(10) = -125,$$

and

$$g(40) = g(20) + \int_{20}^{40} g'(x)\, dx = -125 + \frac{1}{2}(20)(10) = -25.$$

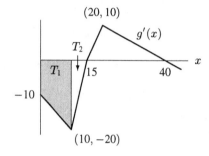

We now find concavity of $g(x)$ in the intervals $[0, 10], [10, 15], [15, 20], [20, 40]$ by checking whether $g'(x)$ increases or decreases in these same intervals. If $g'(x)$ increases, then $g(x)$ is concave up; if $g'(x)$ decreases, then $g(x)$ is concave down. Thus we finally have our graph of $g(x)$:

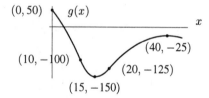

17. Both $F(x)$ and $G(x)$ have roots at $x = 0$ and $x = 4$. Both have a critical point (which is a local maximum) at $x = 2$. However, since the area under $g(x)$ between $x = 0$ and $x = 2$ is larger than the area under $f(x)$ between $x = 0$ and $x = 2$, the y-coordinate of $G(x)$ at 2 will be larger than the y-coordinate of $F(x)$ at 2. See below.

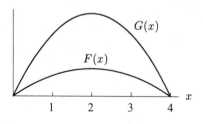

Solutions for Section 6.2

1. $5x$

5. $\sin t$

9. $-\dfrac{1}{2z^2}$

13. $\dfrac{t^4}{4} - \dfrac{t^3}{6} - \dfrac{t^2}{2}$

17. $-\cos 2\theta$

21. $\sin t + \tan t$

25. $f(x) = \frac{1}{4}x$, so $F(x) = \frac{x^2}{8} + C$. $F(0) = 0$ implies that $\frac{1}{8} \cdot 0^2 + C = 0$, so $C = 0$. Thus $F(x) = x^2/8$ is the only possibility.

29. $f(x) = \sin x$, so $F(x) = -\cos x + C$. $F(0) = 0$ implies that $-\cos 0 + C = 0$, so $C = 1$. Thus $F(x) = -\cos x + 1$ is the only possibility.

33. $5e^z + C$

37. $\displaystyle\int \left(t^{3/2} + t^{-3/2}\right) dt = \dfrac{2t^{5/2}}{5} - 2t^{-1/2} + C$

41. $\displaystyle\int \left(y - \dfrac{1}{y}\right)^2 dy = \int \left(y^2 - 2 + \dfrac{1}{y^2}\right) dy = \dfrac{y^3}{3} - 2y - \dfrac{1}{y} + C$

45. $\frac{1}{2}\sin 2x + 2\cos x + C$, since $\frac{d}{dx}(\sin 2x) = 2\cos 2x$.

49. $\displaystyle\int_0^2 \left(\dfrac{x^3}{3} + 2x\right) dx = \left(\dfrac{x^4}{12} + x^2\right)\Big|_0^2 = \dfrac{4}{3} + 4 = 16/3 \approx 5.333$.

53. Since $(\tan x)' = \dfrac{1}{\cos^2 x}$, $\displaystyle\int_0^{\pi/4} \dfrac{1}{\cos^2 x}\, dx = \tan x\Big|_0^{\pi/4} = \tan\dfrac{\pi}{4} - \tan 0 = 1$.

57. The graph is shown in the figure below. Since $\cos\theta \geq \sin\theta$ for $0 \leq \theta \leq \pi/4$, we have

$$\text{Area} = \int_0^{\pi/4} (\cos\theta - \sin\theta)\, d\theta$$
$$= (\sin\theta + \cos\theta)\Big|_0^{\pi/4}$$
$$= \dfrac{1}{\sqrt{2}} + \dfrac{1}{\sqrt{2}} - 1 = \sqrt{2} - 1.$$

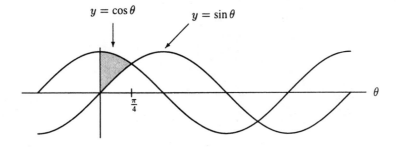

61. (a) The average value of $f(t) = \sin t$ over $0 \le t \le 2\pi$ is given by the formula

$$\text{Average} = \frac{1}{2\pi - 0} \int_0^{2\pi} \sin t \, dt$$

$$= \frac{1}{2\pi}(-\cos t)\Big|_0^{2\pi}$$

$$= \frac{1}{2\pi}(-\cos 2\pi - (-\cos 0)) = 0.$$

We can check this answer by looking at the graph of $\sin t$ below. The area below the curve and above the t-axis over the interval $0 \le t \le \pi$, A_1, is the same as the area above the curve but below the t-axis over the interval $\pi \le t \le 2\pi$, A_2. When we take the integral of $\sin t$ over the entire interval $0 \le t \le 2\pi$, we get $A_1 - A_2 = 0$.

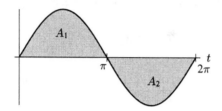

(b) Since

$$\int_0^{\pi} \sin t \, dt = -\cos t\Big|_0^{\pi} = -\cos \pi - (-\cos 0) = -(-1) - (-1) = 2,$$

the average value of $\sin t$ on $0 \le t \le \pi$ is given by

$$\text{Average value} = \frac{1}{\pi} \int_0^{\pi} \sin t \, dt = \frac{2}{\pi}.$$

Solutions for Section 6.3

1. $y = \displaystyle\int (x^3 + 5)\, dx = \dfrac{x^4}{4} + 5x + C$

5. $y = \displaystyle\int (6x^2 + 4x)\, dx = 2x^3 + 2x^2 + C$. If $y(2) = 10$, then $2(2)^3 + 2(2)^2 + C = 10$ and $C = 10 - 16 - 8 = -14$. Thus, $y = 2x^3 + 2x^2 - 14$.

9. We differentiate $y = xe^{-x} + 2$ using the product rule to obtain

$$\frac{dy}{dx} = x\left(e^{-x}(-1)\right) + (1)e^{-x} + 0$$

$$= -xe^{-x} + e^{-x}$$

$$= (1 - x)e^{-x},$$

and so $y = xe^{-x} + 2$ satisfies the differential equation. We now check that $y(0) = 2$:

$$y = xe^{-x} + 2$$

$$y(0) = 0e^0 + 2 = 2.$$

13.   (a)

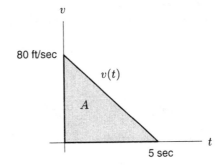

(b)   The total distance is represented by the shaded region A, the area under the graph of $v(t)$.

(c)   The area A, a triangle, is given by

$$A = \frac{1}{2}(\text{base})(\text{height}) = \frac{1}{2}(5\ \text{sec})(80\ \text{ft/sec}) = 200\ \text{ft}.$$

(d)   Using integration and the Fundamental Theorem of Calculus, we have $A = \int_0^5 v(t)\,dt$ or $A = s(5) - s(0)$, where $s(t)$ is an antiderivative of $v(t)$.

   We have that $a(t)$, the acceleration, is constant: $a(t) = k$ for some constant k. Therefore $v(t) = kt + C$ for some constant C. We have $80 = v(0) = k(0) + C = C$, so that $v(t) = kt + 80$. Putting in $t = 5$, $0 = v(5) = (k)(5) + 80$, or $k = -80/5 = -16$.

   Thus $v(t) = -16t + 80$, and an antiderivative for $v(t)$ is $s(t) = -8t^2 + 80t + C$. Since the total distance traveled at $t = 0$ is 0, we have $s(0) = 0$ which means $C = 0$. Finally, $A = \int_0^5 v(t)\,dt = s(5) - s(0) = (-8(5)^2 + (80)(5)) - (-8(0)^2 + (80)(0)) = 200$ ft, which agrees with the previous part.

17.   (a)

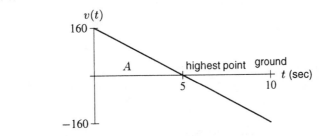

(b)   The highest point is at $t = 5$ seconds. The object hits the ground at $t = 10$ seconds, since by symmetry if the object takes 5 seconds to go up, it takes 5 seconds to come back down.

(c)   The maximum height is the distance traveled when going up, which is represented by the area A of the triangle above the time axis.

$$\text{Area} = \frac{1}{2}(160\ \text{ft/sec})(5\ \text{sec}) = 400\ \text{feet}.$$

(d)   The slope of the line is -32, so $v(t) = -32t + 160$. Antidifferentiating, we get $s(t) = -16t^2 + 160t + s_0$. $s_0 = 0$, so $s(t) = -16t^2 + 160t$. At $t = 5$, $s(t) = -400 + 800 = 400$ ft.

21.   The height of an object above the ground which begins at rest and falls for t seconds is

$$s(t) = -16t^2 + K,$$

where K is the initial height. Here the flower pot falls from 200 ft, so $K = 200$. To see when the pot hits the ground, solve $-16t^2 + 200 = 0$. The solution is

$$t = \sqrt{\frac{200}{16}} \approx 3.54\ \text{seconds}.$$

Now, velocity is given by $s'(t) = v(t) = -32t$. So, the velocity when the pot hits the ground is

$$v(3.54) \approx -113.1\ \text{ft/sec},$$

which is approximately 77 mph downwards.

Solutions for Section 6.4

1.

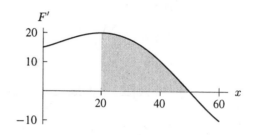

We know that $F(x)$ increases for $x < 50$ because the derivative of F is positive there. See figure above. Similarly, $F(x)$ decreases for $x > 50$. Therefore, the graph of F rises until $x = 50$, and then it begins to fall. Thus, the maximum value attained by F is $F(50)$. To evaluate $F(50)$, we use the Fundamental Theorem:

$$F(50) - F(20) = \int_{20}^{50} F'(x)\,dx,$$

which gives

$$F(50) = F(20) + \int_{20}^{50} F'(x)\,dx = 150 + \int_{20}^{50} F'(x)\,dx.$$

The definite integral equals the area of the shaded region under the graph of F', which is roughly 350. Therefore, the greatest value attained by F is $F(50) \approx 150 + 350 = 500$.

5.

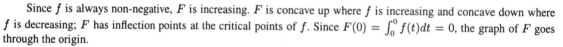

Since f is always non-negative, F is increasing. F is concave up where f is increasing and concave down where f is decreasing; F has inflection points at the critical points of f. Since $F(0) = \int_0^0 f(t)dt = 0$, the graph of F goes through the origin.

9. $(1 + x)^{200}$.

13. Considering $\text{Si}(x^2)$ as the composition of $\text{Si}(u)$ and $u(x) = x^2$, we may apply the chain rule to obtain

$$\frac{d}{dx} = \frac{d(\text{Si}(u))}{du} \cdot \frac{du}{dx}$$
$$= \frac{\sin u}{u} \cdot 2x$$
$$= \frac{2\sin(x^2)}{x}.$$

17. If we let $f(x) = \text{erf}(x)$ and $g(x) = \sqrt{x}$, then we are looking for $\frac{d}{dx}[f(g(x))]$. By the chain rule, this is the same as $g'(x)f'(g(x))$. Since

$$f'(x) = \frac{d}{dx}\left(\frac{2}{\sqrt{\pi}}\int_0^x e^{-t^2}\,dt\right)$$
$$= \frac{2}{\sqrt{\pi}}e^{-x^2}$$

and $g'(x) = \frac{1}{2\sqrt{x}}$, we have

$$f'(g(x)) = \frac{2}{\sqrt{\pi}}e^{-x},$$

and so

$$\frac{d}{dx}[\text{erf}(\sqrt{x})] = \frac{1}{2\sqrt{x}}\frac{2}{\sqrt{\pi}}e^{-x} = \frac{1}{\sqrt{\pi x}}e^{-x}.$$

Solutions for Chapter 6 Review

1. $\frac{5}{2}x^2 + 7x + C$

5. $\tan x + C$

9. $\frac{1}{10}(x+1)^{10} + C$

13. $3\sin x + 7\cos x + C$

17. $\ln|x| - \dfrac{1}{x} - \dfrac{1}{2x^2} + C$

21. Antiderivative $H(r) = 2\dfrac{r^{3/2}}{3/2} + \dfrac{1}{2}\dfrac{r^{1/2}}{1/2} + C = \dfrac{4}{3}r^{3/2} + r^{1/2} + C$

25. $F(z) = e^z + 3z + C$

29. $P(r) = \pi r^2 + C$

33. $\dfrac{1}{2}\sin(t^2) + C$

37. $\dfrac{1}{2}e^{x^2} + C$

41. The graph of $y = c(1-x^2)$ has x-intercepts of $x = \pm 1$. See Figure 6.1. Since it is symmetric about the y-axis, we have

$$\text{Area} = \int_{-1}^{1} c(1-x^2)\,dx = 2c\int_{0}^{1}(1-x^2)\,dx$$

$$= 2c\left(x - \frac{x^3}{3}\right)\bigg|_{0}^{1} = \frac{4c}{3}.$$

We want the area to be 1, so

$$\frac{4c}{3} = 1, \quad \text{giving} \quad c = \frac{3}{4}.$$

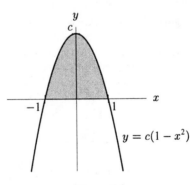

Figure 6.1

45.

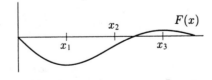

49.   (a)   Since 6 sec = 1/10 min,

$$\text{Angular acceleration } = \frac{2500 - 1100}{1/10} = 14{,}000 \text{ revs/min}^2.$$

(b)   We know angular acceleration is the derivative of angular velocity. Since

$$\text{Angular acceleration } = 14{,}000,$$

we have

$$\text{Angular velocity } = 14{,}000t + C.$$

Measuring time from the moment at which the angular velocity is 1100 revs/min, we have $C = 1100$. Thus,

$$\text{Angular velocity } = 14{,}000t + 1100.$$

Thus the total number of revolutions performed during the period from $t = 0$ to $t = 1/10$ min is given by

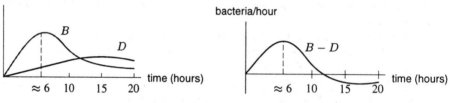

$$\begin{array}{l}\text{Number of} \\ \text{revolutions}\end{array} = \int_0^{1/10} (14000t + 1100)dt = 7000t^2 + 1100t \Big|_0^{1/10} = 180 \text{ revolutions.}$$

53.   (a)   In the beginning, both birth and death rates are small; this is consistent with a very small population. Both rates begin climbing, the birth rate faster than the death rate, which is consistent with a growing population. The birth rate is then high, but it begins to decrease as the population increases.

(b)

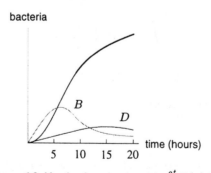

Figure 6.2: Difference between B and D is greatest at $t \approx 6$

The bacteria population is growing most quickly when $B - D$, the rate of change of population, is maximal; that happens when B is farthest above D, which is at a point where the slopes of both graphs are equal. That point is $t \approx 6$ hours.

(c)   Total number born by time t is the area under the B graph from $t = 0$ up to time t. See Figure 6.3.

Figure 6.3: Number born by time t is $\int_0^t B(x)\,dx$

Total number alive at time t is the number born minus the number that have died, which is the area under the B graph minus the area under the D graph, up to time t. See Figure 6.4.

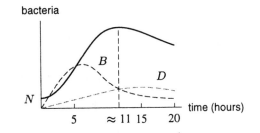

Figure 6.4: Number alive at time t is $\int_0^t (B(x) - D(x))\, dx$

From Figure 6.4, we see that the population is at a maximum when $B = D$, that is, after about 11 hours. This stands to reason, because $B - D$ is the rate of change of population, so population is maximized when $B - D = 0$, that is, when $B = D$.

Solutions to Problems on the Equations of Motion

1. The velocity as a function of time is given by: $v = v_0 + at$. Since the object starts from rest, $v_0 = 0$, and the velocity is just the acceleration times time: $v = -32t$. Integrating this, we get position as a function of time: $y = -16t^2 + y_0$, where the last term, y_0, is the initial position at the top of the tower, so $y_0 = 400$ feet. Thus we have a function giving position as a function of time: $y = -16t^2 + 400$.

    To find at what time the object hits the ground, we find t when $y = 0$. We solve $0 = -16t^2 + 400$ for t, getting $t^2 = 400/16 = 25$, so $t = 5$. Therefore the object hits the ground after 5 seconds. At this time it is moving with a velocity $v = -32(5) = -160$ feet/second.

5. Let the acceleration due to gravity equal $-k$ meters/sec^2, for some positive constant k, and suppose the object falls from an initial height of $s(0)$ meters. We have $a(t) = dv/dt = -k$, so that

$$v(t) = -kt + v_0.$$

Since the initial velocity is zero, we have

$$v(0) = -k(0) + v_0 = 0,$$

which means $v_0 = 0$. Our formula becomes

$$v(t) = \frac{ds}{dt} = -kt.$$

This means

$$s(t) = \frac{-kt^2}{2} + s_0.$$

Since

$$s(0) = \frac{-k(0)^2}{2} + s_0,$$

we have $s_0 = s(0)$, and our formula becomes

$$s(t) = \frac{-kt^2}{2} + s(0).$$

Suppose that the object falls for t seconds. Assuming it hasn't hit the ground, its height is

$$s(t) = \frac{-kt^2}{2} + s(0),$$

so that the distance traveled is

$$s(0) - s(t) = \frac{kt^2}{2} \text{ meters},$$

which is proportional to t^2.

CHAPTER SEVEN

Solutions for Section 7.1

1. (a) $\frac{d}{dx}\sin(x^2+1)=2x\cos(x^2+1)$; $\frac{d}{dx}\sin(x^3+1)=3x^2\cos(x^3+1)$
    (b) (i) $\frac{1}{2}\sin(x^2+1)+C$ (ii) $\frac{1}{3}\sin(x^3+1)+C$
    (c) (i) $-\frac{1}{2}\cos(x^2+1)+C$ (ii) $-\frac{1}{3}\cos(x^3+1)+C$

5. Make the substitution $w=t^3$, $dw=3t^2\,dt$. The general antiderivative is $\int 12t^2\cos(t^3)\,dt=4\sin(t^3)+C$.

9. Make the substitution $w=x^2+1$, $dw=2x\,dx$. We have
$$\int\frac{x}{x^2+1}dx=\frac{1}{2}\int\frac{dw}{w}=\frac{1}{2}\ln|w|+C=\frac{1}{2}\ln(x^2+1)+C.$$
(Notice that since $x^2+1\geq 0$, $|x^2+1|=x^2+1$.)

13. We use the substitution $w=x^2-4$, $dw=2x\,dx$.
$$\int x(x^2-4)^{\frac{7}{2}}\,dx=\frac{1}{2}\int(x^2-4)^{\frac{7}{2}}(2x\,dx)=\frac{1}{2}\int w^{\frac{7}{2}}\,dw$$
$$=\frac{1}{2}(\frac{2}{9}w^{\frac{9}{2}})+C=\frac{1}{9}(x^2-4)^{\frac{9}{2}}+C.$$
Check: $\frac{d}{dx}[\frac{1}{9}(x^2-4)^{\frac{9}{2}}+C]=\frac{1}{9}\left[\frac{9}{2}(x^2-4)^{\frac{7}{2}}\right]2x=x(x^2-4)^{\frac{7}{2}}.$

17. We use the substitution $w=2t-7$, $dw=2\,dt$.
$$\int(2t-7)^{73}\,dt=\frac{1}{2}\int w^{73}\,dw=\frac{1}{(2)(74)}w^{74}+C=\frac{1}{148}(2t-7)^{74}+C.$$
Check: $\frac{d}{dt}\left[\frac{1}{148}(2t-7)^{74}+C\right]=\frac{74}{148}(2t-7)^{73}(2)=(2t-7)^{73}.$

21. We use the substitution $w=-x^2$, $dw=-2x\,dx$.
$$\int xe^{-x^2}\,dx=-\frac{1}{2}\int e^{-x^2}(-2x\,dx)=-\frac{1}{2}\int e^w\,dw$$
$$=-\frac{1}{2}e^w+C=-\frac{1}{2}e^{-x^2}+C.$$
Check: $\frac{d}{dx}(-\frac{1}{2}e^{-x^2}+C)=(-2x)(-\frac{1}{2}e^{-x^2})=xe^{-x^2}.$

25. We use the substitution $w=\sin\alpha$, $dw=\cos\alpha\,d\alpha$.
$$\int\sin^3\alpha\cos\alpha\,d\alpha=\int w^3\,dw=\frac{w^4}{4}+C=\frac{\sin^4\alpha}{4}+C.$$
Check: $\frac{d}{d\alpha}\left(\frac{\sin^4\alpha}{4}+C\right)=\frac{1}{4}\cdot 4\sin^3\alpha\cdot\cos\alpha=\sin^3\alpha\cos\alpha.$

29. We use the substitution $w=\cos 2x$, $dw=-2\sin 2x\,dx$.
$$\int\tan 2x\,dx=\int\frac{\sin 2x}{\cos 2x}\,dx=-\frac{1}{2}\int\frac{dw}{w}$$
$$=-\frac{1}{2}\ln|w|+C=-\frac{1}{2}\ln|\cos 2x|+C.$$

Check:

$$\frac{d}{dx}\left[-\frac{1}{2}\ln|\cos 2x| + C\right] = -\frac{1}{2} \cdot \frac{1}{\cos 2x} \cdot -2\sin 2x$$

$$= \frac{\sin 2x}{\cos 2x} = \tan 2x.$$

33.   We use the substitution $w = 2 + e^x$, $dw = e^x \, dx$.

$$\int \frac{e^x}{2 + e^x} \, dx = \int \frac{dw}{w} = \ln|w| + C = \ln(2 + e^x) + C.$$

(We can drop the absolute value signs since $2 + e^x \geq 0$ for all x.)

Check: $\dfrac{d}{dx}[\ln(2 + e^x) + C] = \dfrac{1}{2 + e^x} \cdot e^x = \dfrac{e^x}{2 + e^x}.$

37.   We use the substitution $w = 1 + 3t^2$, $dw = 6t \, dt$.

$$\int \frac{t}{1 + 3t^2} \, dt = \int \frac{1}{w}(\frac{1}{6}\,dw) = \frac{1}{6}\ln|w| + C = \frac{1}{6}\ln(1 + 3t^2) + C.$$

(We can drop the absolute value signs since $1 + 3t^2 > 0$ for all t).

Check: $\dfrac{d}{dt}\left[\dfrac{1}{6}\ln(1 + 3t^2) + C\right] = \dfrac{1}{6}\dfrac{1}{1 + 3t^2}(6t) = \dfrac{t}{1 + 3t^2}.$

41.   We use the substitution $w = x^2 + x$, $dw = (2x + 1)\,dx$.

$$\int (2x + 1)e^{x^2}e^x \, dx = \int (2x + 1)e^{x^2 + x} \, dx = \int e^w \, dw$$

$$= e^w + C = e^{x^2 + x} + C.$$

Check: $\dfrac{d}{dx}(e^{x^2 + x} + C) = e^{x^2 + x} \cdot (2x + 1) = (2x + 1)e^{x^2}e^x.$

45.   Since $v = \dfrac{dh}{dt}$, it follows that $h(t) = \displaystyle\int v(t)\,dt$ and $h(0) = h_0$. Since

$$v(t) = \frac{mg}{k}\left(1 - e^{-\frac{k}{m}t}\right) = \frac{mg}{k} - \frac{mg}{k}e^{-\frac{k}{m}t},$$

we have

$$h(t) = \int v(t) \, dt = \frac{mg}{k}\int dt - \frac{mg}{k}\int e^{-\frac{k}{m}t} \, dt.$$

The first integral is simply $\dfrac{mg}{k}t + C$. To evaluate the second integral, make the substitution $w = -\frac{k}{m}t$. Then

$$dw = -\frac{k}{m}\,dt,$$

so

$$\int e^{-\frac{k}{m}t} \, dt = \int e^w \left(-\frac{m}{k}\right) dw = -\frac{m}{k}e^w + C = -\frac{m}{k}e^{-\frac{k}{m}t} + C.$$

Thus

$$h(t) = \int v\,dt = \frac{mg}{k}t - \frac{mg}{k}\left(-\frac{m}{k}e^{-\frac{k}{m}t}\right) + C$$

$$= \frac{mg}{k}t + \frac{m^2 g}{k^2}e^{-\frac{k}{m}t} + C.$$

Since $h(0) = h_0$,

$$h_0 = \frac{mg}{k} \cdot 0 + \frac{m^2 g}{k^2}e^0 + C;$$

$$C = h_0 - \frac{m^2 g}{k^2}.$$

Thus

$$h(t) = \frac{mg}{k}t + \frac{m^2g}{k^2}e^{-\frac{k}{m}t} - \frac{m^2g}{k^2} + h_0$$

$$h(t) = \frac{mg}{k}t - \frac{m^2g}{k^2}\left(1 - e^{-\frac{k}{m}t}\right) + h_0.$$

Solutions for Section 7.2

1. (a) We substitute $w = 1 + x^2, dw = 2x \, dx$.

$$\int_{x=0}^{x=1} \frac{x}{1+x^2} \, dx = \frac{1}{2} \int_{w=1}^{w=2} \frac{1}{w} \, dw = \frac{1}{2} \ln|w| \Big|_1^2 = \frac{1}{2} \ln 2.$$

(b) We substitute $w = \cos x, dw = -\sin x \, dx$.

$$\int_{x=0}^{x=\frac{\pi}{4}} \frac{\sin x}{\cos x} \, dx = -\int_{w=1}^{w=\sqrt{2}/2} \frac{1}{w} \, dw$$

$$= -\ln|w| \Big|_1^{\sqrt{2}/2} = -\ln\frac{\sqrt{2}}{2} = \frac{1}{2}\ln 2.$$

5. $\displaystyle\int_1^2 2xe^{x^2} \, dx = e^{x^2} \Big|_1^2 = e^{2^2} - e^{1^2} = e^4 - e = e(e^3 - 1)$

9. We substitute $w = 1 + x^2$. Then $dw = 2x \, dx$.

$$\int_{x=0}^{x=2} \frac{x}{(1+x^2)^2} \, dx = \int_{w=1}^{w=5} \frac{1}{w^2}\left(\frac{1}{2} \, dw\right) = -\frac{1}{2}\left(\frac{1}{w}\right) \Big|_1^5 = \frac{2}{5}.$$

13. $\displaystyle\int_1^3 \frac{1}{x} \, dx = \ln x \Big|_1^3 = \ln 3.$

17. Substitute $w = 1 + x^2, dw = 2x \, dx$. Then $x \, dx = \frac{1}{2} \, dw$, and

$$\int_{x=0}^{x=1} x(1+x^2)^{20} \, dx = \frac{1}{2}\int_{w=1}^{w=2} w^{20} \, dw = \frac{w^{21}}{42} \Big|_1^2 = \frac{299593}{6} = 49932\frac{1}{6}.$$

21.

$$\int_1^2 \frac{x^2+1}{x} \, dx = \int_1^2 \left(x + \frac{1}{x}\right) dx = \left(\frac{x^2}{2} + \ln|x|\right) \Big|_1^2 = \frac{3}{2} + \ln 2.$$

25. Let $w = x^2, dw = 2x \, dx$. When $x = 0$, $w = 0$, and when $x = \frac{1}{\sqrt{2}}$, $w = \frac{1}{2}$. Then

$$\int_0^{\frac{1}{\sqrt{2}}} \frac{x \, dx}{\sqrt{1-x^4}} = \int_0^{\frac{1}{2}} \frac{\frac{1}{2} \, dw}{\sqrt{1-w^2}} = \frac{1}{2} \arcsin w \Big|_0^{\frac{1}{2}} = \frac{1}{2}\left(\arcsin\frac{1}{2} - \arcsin 0\right) = \frac{\pi}{12}.$$

29. The substitution $w = \ln x, dw = \frac{1}{x} \, dx$ transforms the first integral into $\int w \, dw$, which is just a respelling of the integral $\int x \, dx$.

33.

$$\int \frac{dx}{x^2 + 4x + 5} = \int \frac{dx}{(x + 2)^2 + 1}.$$

We make the substitution $\tan\theta = x + 2$. Then $dx = \frac{1}{\cos^2\theta}\,d\theta$.

$$\int \frac{dx}{(x + 2)^2 + 1} = \int \frac{d\theta}{\cos^2\theta(\tan^2\theta + 1)}$$
$$= \int \frac{d\theta}{\cos^2\theta(\frac{\sin^2\theta}{\cos^2\theta} + 1)}$$
$$= \int \frac{d\theta}{\sin^2\theta + \cos^2\theta}$$
$$= \int d\theta = \theta + C$$

But since $\tan\theta = x + 2$, $\theta = \arctan(x + 2)$, and so $\theta + C = \arctan(x + 2) + C$.

37. (a) The Fundamental Theorem gives

$$\int_{-\pi}^{\pi} \cos^2\theta \sin\theta\, d\theta = -\frac{\cos^3\theta}{3}\Big|_{-\pi}^{\pi} = \frac{-(-1)^3}{3} - \frac{-(-1)^3}{3} = 0.$$

This agrees with the fact that the function $f(\theta) = \cos^2\theta \sin\theta$ is odd and the interval of integration is centered at $x = 0$, thus we must get 0 for the definite integral.

(b) The area is given by

$$\text{Area} = \int_0^{\pi} \cos^2\theta \sin\theta\, d\theta = -\frac{\cos^3\theta}{3}\Big|_0^{\pi} = \frac{-(-1)^3}{3} - \frac{-(1)^3}{3} = \frac{2}{3}.$$

41. (a) For a 40,000-word novel,

$$\text{Payment} = \int_0^{40,000} f(w)\, dw$$
$$= \int_0^{2000} \left(\frac{1}{2} + \frac{w}{2000}\right) dw + \int_{2000}^{20,000} \frac{3}{2}\, dw + \int_{20,000}^{40,000} \frac{3}{2}e^{20 - \frac{w}{1000}}\, dw$$
$$= \left(\frac{1}{2}w + \frac{w^2}{4000}\right)\Big|_0^{2000} + \left(\frac{3}{2}w\right)\Big|_{2000}^{20,000} + \left(\frac{-3000}{2}e^{20 - \frac{w}{1000}}\right)\Big|_{20,000}^{40,000}$$
$$= 2000 + 27{,}000 + 1500 - 1500e^{-20}$$
$$\approx 30{,}500 \text{ pennies.}$$

(b) Notice from part (a) that the payment for a 40,000-word novel is 2000 pennies (or a penny a word) for the first 2000 pages, 27,000 pennies (or one and a half pennies a word) for the next 18,000 pages, and it is only 1,500 pennies for the final 20,000 words. Thus, the payment is greater for two 15,000-word novels, since the function discourages long novels.

Solutions for Section 7.3

1. Let $u = \arctan x$, $v' = 1$. Then $v = x$ and $u' = \frac{1}{1 + x^2}$. Integrating by parts, we get:

$$\int 1 \cdot \arctan x\, dx = x \cdot \arctan x - \int x \cdot \frac{1}{1 + x^2}\, dx.$$

To compute the second integral use the substitution, $z = 1 + x^2$.

$$\int \frac{x}{1+x^2}\, dx = \frac{1}{2}\int \frac{dz}{z} = \frac{1}{2}\ln|z| + C = \frac{1}{2}\ln(1+x^2) + C.$$

Thus,

$$\int \arctan x\, dx = x \cdot \arctan x - \frac{1}{2}\ln(1+x^2) + C.$$

5. Let $u = t$, $v' = \sin t$. Thus, $v = -\cos t$ and $u' = 1$. With this choice of u and v, integration by parts gives:

$$\int t\sin t\, dt = -t\cos t - \int (-\cos t)\, dt$$
$$= -t\cos t + \sin t + C.$$

9. Let $u = z$, $v' = e^{-z}$. Thus $v = -e^{-z}$ and $u' = 1$. Integration by parts gives:

$$\int ze^{-z}\, dz = -ze^{-z} - \int (-e^{-z})\, dz$$
$$= -ze^{-z} - e^{-z} + C$$
$$= -(z+1)e^{-z} + C.$$

13. Let $u = \theta + 1$ and $v' = \sin(\theta + 1)$, so $u' = 1$ and $v = -\cos(\theta + 1)$.

$$\int (\theta + 1)\sin(\theta + 1)\, d\theta = -(\theta + 1)\cos(\theta + 1) + \int \cos(\theta + 1)\, d\theta$$
$$= -(\theta + 1)\cos(\theta + 1) + \sin(\theta + 1) + C.$$

17. Let $u = y$ and $v' = (y+3)^{1/2}$, so $u' = 1$ and $v = \frac{2}{3}(y+3)^{3/2}$.
$$\int y\sqrt{y+3}\, dy = \frac{2}{3}y(y+3)^{3/2} - \int \frac{2}{3}(y+3)^{3/2}\, dy = \frac{2}{3}y(y+3)^{3/2} - \frac{4}{15}(y+3)^{5/2} + C.$$

21. Let $u = (\ln t)^2$ and $v' = 1$, so $u' = \dfrac{2\ln t}{t}$ and $v = t$. Then

$$\int (\ln t)^2\, dt = t(\ln t)^2 - 2\int \ln t\, dt = t(\ln t)^2 - 2t\ln t + 2t + C.$$

(We use the fact that $\displaystyle\int \ln x\, dx = x\ln x - x + C$, a result which can be derived using integration by parts.)

25. This integral can first be simplified by making the substitution $w = x^2$, $dw = 2x\, dx$. Then

$$\int x\arctan x^2\, dx = \frac{1}{2}\int \arctan w\, dw.$$

To evaluate $\int \arctan w\, dw$, we'll use integration by parts. Let $u = \arctan w$ and $v' = 1$, so $u' = \frac{1}{1+w^2}$ and $v = w$. Then

$$\int \arctan w\, dw = w\arctan w - \int \frac{w}{1+w^2}\, dw = w\arctan w - \frac{1}{2}\ln|1+w^2| + C.$$

Since $1 + w^2$ is never negative, we can drop the absolute value signs. Thus, we have

$$\int x\arctan x^2\, dx = \frac{1}{2}\left(x^2\arctan x^2 - \frac{1}{2}\ln(1+(x^2)^2) + C\right)$$
$$= \frac{1}{2}x^2\arctan x^2 - \frac{1}{4}\ln(1+x^4) + C.$$

29. $\displaystyle\int_3^5 x \cos x \, dx = (\cos x + x \sin x)\Big|_3^5 = \cos 5 + 5 \sin 5 - \cos 3 - 3 \sin 3 \approx -3.944.$

33. $\displaystyle\int_0^5 \ln(1+t) \, dt = \big((1+t)\ln(1+t) - (1+t)\big)\Big|_0^5 = 6\ln 6 - 5 \approx 5.751.$

37. From integration by parts in Problem 12, we obtain

$$\int \sin^2 \theta \, d\theta = -\frac{1}{2} \sin \theta \cos \theta + \frac{1}{2}\theta + C.$$

Using the identity given in the book, we have

$$\int \sin^2 \theta \, d\theta = \int \frac{1 - \cos 2\theta}{2} \, d\theta = \frac{1}{2}\theta - \frac{1}{4} \sin 2\theta + C.$$

Although the answers differ in form, they are really the same, since (by one of the standard double angle formulas) $-\frac{1}{4} \sin 2\theta = -\frac{1}{4}(2 \sin \theta \cos \theta) = -\frac{1}{2} \sin \theta \cos \theta.$

41. We integrate by parts. Since in Problem 39 we found that $\int e^x \sin x \, dx = \frac{1}{2}e^x(\sin x - \cos x)$, we let $u = x$ and $v' = e^x \sin x$, so $u' = 1$ and $v = \frac{1}{2}e^x(\sin x - \cos x)$.

$$\text{Then } \int xe^x \sin x \, dx = \frac{1}{2}xe^x(\sin x - \cos x) - \frac{1}{2}\int e^x(\sin x - \cos x)\, dx$$

$$= \frac{1}{2}xe^x(\sin x - \cos x) - \frac{1}{2}\int e^x \sin x \, dx + \frac{1}{2}\int e^x \cos x \, dx.$$

Using Problems 39 and 40, we see that this equals

$$\frac{1}{2}xe^x(\sin x - \cos x) - \frac{1}{4}e^x(\sin x - \cos x) + \frac{1}{4}e^x(\sin x + \cos x) + C$$

$$= \frac{1}{2}xe^x(\sin x - \cos x) + \frac{1}{2}e^x \cos x + C.$$

45. We integrate by parts. Let $u = x^n$ and $v' = \sin ax$, so $u' = nx^{n-1}$ and $v = -\frac{1}{a}\cos ax$.

$$\text{Then } \int x^n \sin ax \, dx = -\frac{1}{a}x^n \cos ax - \int (nx^{n-1})(-\frac{1}{a}\cos ax)\, dx$$

$$= -\frac{1}{a}x^n \cos ax + \frac{n}{a}\int x^{n-1} \cos ax \, dx.$$

49. (a) We have

$$F(a) = \int_0^a x^2 e^{-x} \, dx$$

$$= -x^2 e^{-x}\Big|_0^a + \int_0^a 2xe^{-x}\, dx$$

$$= (-x^2 e^{-x} - 2xe^{-x})\Big|_0^a + 2\int_0^a e^{-x}\, dx$$

$$= (-x^2 e^{-x} - 2xe^{-x} - 2e^{-x})\Big|_0^a$$

$$= -a^2 e^{-a} - 2ae^{-a} - 2e^{-a} + 2.$$

(b) $F(a)$ is increasing because $x^2 e^{-x}$ is positive, so as a increases, the area under the curve from 0 to a also increases and thus the integral increases.

(c) We have $F'(a) = a^2 e^{-a}$, so

$$F''(a) = 2ae^{-a} - a^2 e^{-a} = a(2 - a)e^{-a}.$$

We see that $F''(a) > 0$ for $0 < a < 2$, so F is concave up on this interval.

53.   (a)   We want to compute C_1, with $C_1 > 0$, such that

$$\int_0^1 \left(\Psi_1(x)\right)^2 dx = \int_0^1 \left(C_1 \sin(\pi x)\right)^2 dx = C_1^2 \int_0^1 \sin^2(\pi x)\, dx = 1.$$

We use integration by parts with $u = v' = \sin(\pi x)$.
So $u' = \pi \cos(\pi x)$ and $v = -\frac{1}{\pi} \cos(\pi x)$. Thus

$$\int_0^1 \sin^2(\pi x)\, dx = -\frac{1}{\pi} \sin(\pi x) \cos(\pi x)\Big|_0^1 + \int_0^1 \cos^2(\pi x)\, dx$$

$$= -\frac{1}{\pi} \sin(\pi x) \cos(\pi x)\Big|_0^1 + \int_0^1 (1 - \sin^2(\pi x))\, dx.$$

Moving $\int_0^1 \sin^2(\pi x)\, dx$ from the right side to the left side of the equation and solving, we get

$$2\int_0^1 \sin^2(\pi x)\, dx = -\frac{1}{\pi} \sin(\pi x) \cos(\pi x)\Big|_0^1 + \int_0^1 1\, dx = 0 + 1 = 1,$$

so

$$\int_0^1 \sin^2(\pi x)\, dx = \frac{1}{2}.$$

Thus, we have

$$\int_0^1 \left(\Psi_1(x)\right)^2 dx = C_1^2 \int_0^1 \sin^2(\pi x)\, dx = \frac{C_1^2}{2}.$$

So, to normalize Ψ_1, we take $C_1 > 0$ such that

$$\frac{C_1^2}{2} = 1 \quad \text{so} \quad C_1 = \sqrt{2}.$$

   (b)   To normalize Ψ_n, we want to compute C_n, with $C_n > 0$, such that

$$\int_0^1 \left(\Psi_n(x)\right)^2 dx = C_n^2 \int_0^1 \sin^2(n\pi x)\, dx = 1.$$

The solution to part (a) shows us that

$$\int \sin^2(\pi t)\, dt = -\frac{1}{2\pi} \sin(\pi t) \cos(\pi t) + \frac{1}{2} \int 1\, dt.$$

In the integral for Ψ_n, we make the substitution $t = nx$, so $dx = \frac{1}{n} dt$. Since $t = 0$ when $x = 0$ and $t = n$ when $x = 1$, we have

$$\int_0^1 \sin^2(n\pi x)\, dx = \frac{1}{n} \int_0^n \sin^2(\pi t)\, dt$$

$$= \frac{1}{n} \left(-\frac{1}{2\pi} \sin(\pi t) \cos(\pi t)\Big|_0^n + \frac{1}{2} \int_0^n 1\, dt \right)$$

$$= \frac{1}{n} \left(0 + \frac{n}{2} \right) = \frac{1}{2}.$$

Thus, we have

$$\int_0^1 \left(\Psi_n(x)\right)^2 dx = C_n^2 \int_0^1 \sin^2(n\pi x)\, dx = \frac{C_n^2}{2}.$$

So to normalize Ψ_n, we take C_n such that

$$\frac{C_n^2}{2} = 1 \quad \text{so} \quad C_n = \sqrt{2}.$$

Solutions for Section 7.4

1. $\frac{1}{10}e^{(-3\theta)}(-3\cos\theta + \sin\theta) + C.$
    (Let $a = -3, b = 1$ in II-9.)

5. $\frac{1}{\sqrt{3}} \arctan \frac{y}{\sqrt{3}} + C.$
    (Let $a = \sqrt{3}$ in V-24).

9. $\frac{3}{16}\cos 3\theta \sin 5\theta - \frac{5}{16}\sin 3\theta \cos 5\theta + C.$
    (Let $a = 3, b = 5$ in II-10.)

13. Substitute $w = 5u$, $dw = 5\,du$. Then

$$\int u^5 \ln(5u)\,du = \frac{1}{5^6}\int w^5 \ln w\,dw$$
$$= \frac{1}{5^6}(\frac{1}{6}w^6 \ln w - \frac{1}{36}w^6 + C)$$
$$= \frac{1}{6}u^6 \ln 5u - \frac{1}{36}u^6 + C.$$

Or use $\ln 5u = \ln 5 + \ln u$.

$$\int u^5 \ln 5u\,du = \ln 5 \int u^5\,du + \int u^5 \ln u\,du$$
$$= \frac{u^6}{6}\ln 5 + \frac{1}{6}u^6 \ln u - \frac{1}{36}u^6 + C \quad \text{(using III-13)}$$
$$= \frac{u^6}{6}\ln 5u - \frac{1}{36}u^6 + C.$$

17.

$$\int y^2 \sin 2y\,dy = -\frac{1}{2}y^2 \cos 2y + \frac{1}{4}(2y)\sin 2y + \frac{1}{8}(2)\cos 2y + C$$
$$= -\frac{1}{2}y^2 \cos 2y + \frac{1}{2}y \sin 2y + \frac{1}{4}\cos 2y + C.$$

(Use $a = 2, p(y) = y^2$ in III-15.)

21. Substitute $w = 7x$, $dw = 7\,dx$. Then use IV-21.

$$\int \frac{1}{\cos^4 7x}\,dx = \frac{1}{7}\int \frac{1}{\cos^4 w}\,dw = \frac{1}{7}\left[\frac{1}{3}\frac{\sin w}{\cos^3 w} + \frac{2}{3}\int \frac{1}{\cos^2 w}\,dw\right]$$
$$= \frac{1}{21}\frac{\sin w}{\cos^3 w} + \frac{2}{21}\left[\frac{\sin w}{\cos w} + C\right]$$
$$= \frac{1}{21}\frac{\tan w}{\cos^2 w} + \frac{2}{21}\tan w + C$$
$$= \frac{1}{21}\frac{\tan 7x}{\cos^2 7x} + \frac{2}{21}\tan 7x + C.$$

25.

$$\int \frac{dy}{4 - y^2} = -\int \frac{dy}{(y+2)(y-2)} = -\frac{1}{4}(\ln|y - 2| - \ln|y + 2|) + C.$$

(Let $a = 2, b = -2$ in V-26.)

29.

$$\int \sin^3 3\theta \cos^2 3\theta \, d\theta = \int (\sin 3\theta)(\cos^2 3\theta)(1 - \cos^2 3\theta) \, d\theta$$

$$= \int \sin 3\theta (\cos^2 3\theta - \cos^4 3\theta) \, d\theta.$$

Using an extension of the tip given in rule IV-23, we let $w = \cos 3\theta$, $dw = -3\sin 3\theta \, d\theta$.

$$\int \sin 3\theta (\cos^2 3\theta - \cos^4 3\theta) \, d\theta = -\frac{1}{3} \int (w^2 - w^4) \, dw$$

$$= -\frac{1}{3}\left(\frac{w^3}{3} - \frac{w^5}{5}\right) + C$$

$$= -\frac{1}{9}(\cos^3 3\theta) + \frac{1}{15}(\cos^5 3\theta) + C.$$

33. (a) $\dfrac{2}{x} + \dfrac{1}{x+3} = \dfrac{2(x+3)}{x(x+3)} + \dfrac{x}{x(x+3)} = \dfrac{3x+6}{x^2+3x}$. Thus

$$\int \frac{3x+6}{x^2+3x} \, dx = \int \left(\frac{2}{x} + \frac{1}{x+3}\right) dx = 2\ln|x| + \ln|x+3| + C.$$

(b) Let $a = 0, b = -3, c = 3$, and $d = 6$ in V-27.

$$\int \frac{3x+6}{x^2+3x} \, dx = \int \frac{3x+6}{x(x+3)} \, dx$$

$$= \frac{1}{3}(6\ln|x| + 3\ln|x+3|) + C = 2\ln|x| + \ln|x+3| + C.$$

37. We write

$$\frac{1}{3P - 3P^2} = \frac{A}{3P} + \frac{B}{1-P},$$

multiply through by $3P(1-P)$, and then solve for A and B, getting $A = 1$ and $B = 1/3$. So

$$\int \frac{dP}{3P - 3P^2} = \int \left(\frac{1}{3P} + \frac{1}{3(1-P)}\right) dP = \frac{1}{3}\int \frac{dP}{P} + \frac{1}{3}\int \frac{dP}{1-P}$$

$$= \frac{1}{3}\ln|P| - \frac{1}{3}\ln|1-P| + C = \frac{1}{3}\ln\left|\frac{P}{1-P}\right| + C.$$

41. We want to calculate

$$\int_0^1 C_n \sin(n\pi x) \cdot C_m \sin(m\pi x) \, dx.$$

We use II-11 from the table of integrals with $a = n\pi, b = m\pi$. Since $n \neq m$, we see that

$$\int_0^1 \Psi_n(x) \cdot \Psi_m(x) \, dx = C_n C_m \int_0^1 \sin(n\pi x)\sin(m\pi x) \, dx$$

$$= \frac{C_n C_m}{m^2\pi^2 - n^2\pi^2}\left(n\pi\cos(n\pi x)\sin(m\pi x) - m\pi\sin(n\pi x)\cos(m\pi x)\right)\Big|_0^1$$

$$= \frac{C_n C_m}{(m^2 - n^2)\pi^2}\left(n\pi\cos(n\pi)\sin(m\pi) - m\pi\sin(n\pi)\cos(m\pi)\right.$$

$$\left. -n\pi\cos(0)\sin(0) + m\pi\sin(0)\cos(0)\right)$$

$$= 0$$

since $\sin(0) = \sin(n\pi) = \sin(m\pi) = 0$.

Solutions for Section 7.5

1. (a)
$$\text{LEFT}(2) = 2 \cdot f(0) + 2 \cdot f(2)$$
$$= 2 \cdot 1 + 2 \cdot 5$$
$$= 12$$
$$\text{RIGHT}(2) = 2 \cdot f(2) + 2 \cdot f(4)$$
$$= 2 \cdot 5 + 2 \cdot 17$$
$$= 44$$

(b)

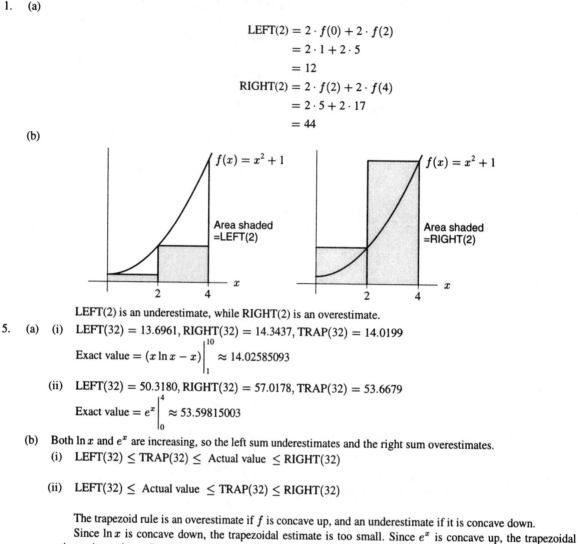

LEFT(2) is an underestimate, while RIGHT(2) is an overestimate.

5. (a) (i)   LEFT(32) = 13.6961, RIGHT(32) = 14.3437, TRAP(32) = 14.0199

Exact value $= (x \ln x - x)\Big|_1^{10} \approx 14.02585093$

(ii)   LEFT(32) = 50.3180, RIGHT(32) = 57.0178, TRAP(32) = 53.6679

Exact value $= e^x \Big|_0^4 \approx 53.59815003$

(b)   Both $\ln x$ and e^x are increasing, so the left sum underestimates and the right sum overestimates.

(i)   LEFT(32) $\leq$ TRAP(32) $\leq$ Actual value $\leq$ RIGHT(32)

(ii)   LEFT(32) $\leq$ Actual value $\leq$ TRAP(32) $\leq$ RIGHT(32)

The trapezoid rule is an overestimate if f is concave up, and an underestimate if it is concave down.

Since $\ln x$ is concave down, the trapezoidal estimate is too small. Since e^x is concave up, the trapezoidal estimate is too large. In each case, however, the trapezoidal estimate should be better than the left- or right-hand sums, since it is the average of the two.

9. (a)   TRAP(4) gives probably the best estimate of the integral. We cannot calculate MID(4).
$$\text{LEFT}(4) = 3 \cdot 100 + 3 \cdot 97 + 3 \cdot 90 + 3 \cdot 78 = 1095$$
$$\text{RIGHT}(4) = 3 \cdot 97 + 3 \cdot 90 + 3 \cdot 78 + 3 \cdot 55 = 960$$
$$\text{TRAP}(4) = \frac{1095 + 960}{2} = 1027.5.$$

(b)   Because there are no points of inflection, the graph is either concave down or concave up. By plotting points, we see that it is concave down. So TRAP(4) is an underestimate.

13. $f(x)$ is concave up, so TRAP gives an overestimate and MID gives an underestimate.

17. We approximate the area of the playing field by using Riemann sums. From the data provided,
$$\text{LEFT}(10) = \text{RIGHT}(10) = \text{TRAP}(10) = 89,000 \text{ square feet.}$$

Thus approximately
$$\frac{89,000 \text{ sq. ft.}}{200 \text{ sq. ft./lb.}} = 445 \text{ lbs. of fertilizer}$$

should be necessary.

21.

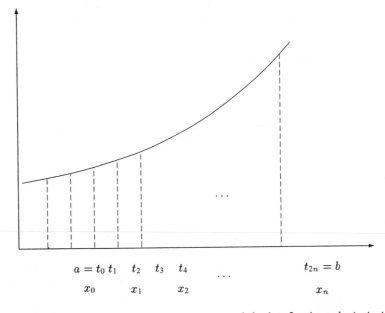

$$a = t_0 \; t_1 \qquad t_2 \qquad t_3 \qquad t_4 \qquad \cdots \qquad t_{2n} = b$$
$$x_0 \qquad\quad x_1 \qquad\quad x_2 \qquad\qquad\qquad x_n$$

Divide the interval $[a, b]$ into n pieces, by $x_0, x_1, x_2, \ldots, x_n$, and also into $2n$ pieces, by $t_0, t_1, t_2, \ldots, t_{2n}$. Then the x's coincide with the even t's, so $x_0 = t_0$, $x_1 = t_2$, $x_2 = t_4$, $\ldots$, $x_n = t_{2n}$ and $\Delta t = \frac{1}{2}\Delta x$.

$$\text{LEFT}(n) = f(x_0)\Delta x + f(x_1)\Delta x + \cdots + f(x_{n-1})\Delta x$$

Since $\text{MID}(n)$ is obtained by evaluating f at the midpoints $t_1, t_3, t_5, \ldots$ of the x intervals, we get

$$\text{MID}(n) = f(t_1)\Delta x + f(t_3)\Delta x + \cdots + f(t_{2n-1})\Delta x$$

Now

$$\text{LEFT}(2n) = f(t_0)\Delta t + f(t_1)\Delta t + f(t_2)\Delta t + \cdots + f(t_{2n-1})\Delta t.$$

Regroup terms, putting all the even t's first, the odd t's last:

$$\text{LEFT}(2n) = f(t_0)\Delta t + f(t_2)\Delta t + \cdots + f(t_{2n-2})\Delta t + f(t_1)\Delta t + f(t_3)\Delta t + \cdots + f(t_{2n-1})\Delta t$$

$$= \underbrace{f(x_0)\frac{\Delta x}{2} + f(x_1)\frac{\Delta x}{2} + \cdots + f(x_{n-1})\frac{\Delta x}{2}}_{\text{LEFT}(n)/2} + \underbrace{f(t_1)\frac{\Delta x}{2} + f(t_3)\frac{\Delta x}{2} + \cdots + f(t_{2n-1})\frac{\Delta x}{2}}_{\text{MID}(n)/2}$$

So

$$\text{LEFT}(2n) = \frac{1}{2}(\text{LEFT}(n) + \text{MID}(n))$$

Solutions for Section 7.6

1. (a) From Problem 2 on page 92, for $\int_0^4 (x^2 + 1)\,dx$, we have MID(2)= 24 and TRAP(2)= 28. Thus,

$$\text{SIMP}(2) = \frac{2\text{MID}(2) + \text{TRAP}(2)}{3}$$
$$= \frac{2(24) + 28}{3}$$
$$= \frac{76}{3}.$$

(b)

$$\int_0^4 (x^2 + 1)\,dx = \left(\frac{x^3}{3} + x\right)\Bigg|_0^4 = \left(\frac{64}{3} + 4\right) - (0 + 0) = \frac{76}{3}$$

(c) Error= 0. Simpson's Rule gives the exact answer.

5.   1.0894. ($n = 10$ intervals, or more)

9.   (a) $\displaystyle\int_0^4 e^x\, dx = e^x \Big|_0^4 = e^4 - e^0 \approx 53.598\ldots.$

     (b)   Computing the sums directly, since $\Delta x = 2$, we have
           LEFT(2)$= 2 \cdot e^0 + 2 \cdot e^2 \approx 2(1) + 2(7.389) = 16.778;$ error $= 36.820.$
           RIGHT(2)$= 2 \cdot e^2 + 2 \cdot e^4 \approx 2(7.389) + 2(54.598) = 123.974;$ error $= -70.376.$
           TRAP(2)$= \dfrac{16.778 + 123.974}{2} = 70.376;$ error $= 16.778.$
           MID(2)$= 2 \cdot e^1 + 2 \cdot e^3 \approx 2(2.718) + 2(20.086) = 45.608;$ error $= 7.990.$
           SIMP(2)$= \dfrac{2(45.608) + 70.376}{3} = 53.864;$ error $= -0.266.$

     (c)   Similarly, since $\Delta x = 1$, we have LEFT(4)$= 31.193;$ error $= 22.405$
           RIGHT(4)$= 84.791;$ error $= -31.193$
           TRAP(4)$= 57.992;$ error $= -4.394$
           MID(4)$= 51.428;$ error $= 2.170$
           SIMP(4)$= 53.616;$ error $= -0.018$

     (d)   For LEFT and RIGHT, we expect the error to go down by $1/2$, and this is very roughly what we see. For MID and TRAP, we expect the error to go down by $1/4$, and this is approximately what we see. For SIMP, we expect the error to go down by $1/2^4 = 1/16$, and this is approximately what we see.

13. (a)   Since the time for a left-hand approximation is proportional to its accuracy, getting 8 more digits of accuracy will take a factor of 10^8 more time, or, in other words, will take 3 seconds $\times 10^8$, or 9.5 years.

     (b)   Similarly to part (a), the midpoint rule increases in time with the square root of accuracy, so it will take 3 seconds $\times 10^4$, or 8.33 hours.

     (c)   Simpson's rule increases in time with the fourth root of accuracy, so it will only take 3 seconds $\times 10^2$, or 5 minutes, to get our desired accuracy.

17. (a)   Since MID(n) seems to be decreasing as n increases, we can assume that MID(10) and MID(20) are both overestimates. We know that the error in the midpoint rule is approximately proportional to $1/n^2$. In going from $n = 10$ to $n = 20$, n is multiplied by 2 and so we expect the error to go down roughly by a factor of $1/2^2$, or $1/4$. Therefore, if we let $d = |\text{error}(20)|$, then we have $4d \approx |\text{error}(10)|$.

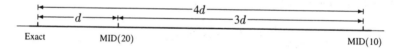

           As we see from the figure above, $3d$ is equal to the difference between MID(10) and MID(20), so
           $$3d = \text{MID}(10) - \text{MID}(20)$$
           $$3d = 5.364 - 4.926$$
           $$d = 0.146.$$
           Since d is the magnitude of the error for MID(20), and since the exact value is less than MID(20), we have
           $$\text{Exact} \approx \text{MID}(20) - d$$
           $$= 4.926 - 0.146 = 4.780.$$
           The exact value of the integral is about 4.780.

     (b)   In part (a), we estimated that the error for MID(20) ≈ 0.146. As n goes from 20 to 60, n is multiplied by 3 and so we expect the error to go down by $1/9$. The error for MID(60) is about $(1/9)0.146 = 0.0162$. Since we estimated the exact value at 4.780, we estimate that MID(60) is about $4.780 + 0.0162 = 4.7962$.

21. (a)   Suppose $q_i(x)$ is the quadratic function approximating $f(x)$ on the subinterval $[x_i, x_{i+1}]$, and m_i is the midpoint of the interval, $m_i = (x_i + x_{i+1})/2$. Then, using the equation in Problem 20, with $a = x_i$ and $b = x_{i+1}$ and $h = \Delta x = x_{i+1} - x_i$:
           $$\int_{x_i}^{x_{i+1}} f(x)dx \approx \int_{x_i}^{x_{i+1}} q_i(x)dx = \frac{\Delta x}{3}\left(\frac{q_i(x_i)}{2} + 2q_i(m_i) + \frac{q_i(x_{i+1})}{2}\right).$$

(b) Summing over all subintervals gives

$$\int_a^b f(x)dx \approx \sum_{i=0}^{n-1} \int_{x_i}^{x_{i+1}} q_i(x)dx = \sum_{i=0}^{n-1} \frac{\Delta x}{3}\left(\frac{q_i(x_i)}{2} + 2q_i(m_i) + \frac{q_i(x_{i+1})}{2}\right).$$

Splitting the sum into two parts:

$$= \frac{2}{3}\sum_{i=0}^{n-1} q_i(m_i)\Delta x + \frac{1}{3}\sum_{i=0}^{n-1} \frac{q_i(x_i) + q_i(x_{i+1})}{2}\Delta x$$

$$= \frac{2}{3}\text{MID}(n) + \frac{1}{3}\text{TRAP}(n)$$

$$= \text{SIMP}(n).$$

Solutions for Section 7.7

1.

$$\int_1^\infty e^{-2x}\,dx = \lim_{b\to\infty}\int_1^b e^{-2x}\,dx = \lim_{b\to\infty} -\frac{e^{-2x}}{2}\Big|_1^b$$

$$= \lim_{b\to\infty}(-e^{-2b}/2 + e^{-2}/2) = 0 + e^{-2}/2 = e^{-2}/2,$$

where the first limit is 0 because $\lim_{x\to\infty} e^{-x} = 0$.

5. First, we note that $1/(z^2 + 25)$ is an even function. Therefore,

$$\int_{-\infty}^\infty \frac{dz}{z^2+25} = \int_{-\infty}^0 \frac{dz}{z^2+25} + \int_0^\infty \frac{dz}{z^2+25} = 2\int_0^\infty \frac{dz}{z^2+25}.$$

We'll now evaluate this improper integral by using a limit:

$$\int_0^\infty \frac{dz}{z^2+25} = \lim_{b\to\infty}\left(\frac{1}{5}\arctan(b/5) - \frac{1}{5}\arctan(0)\right) = \frac{1}{5}\cdot\frac{\pi}{2} = \frac{\pi}{10}.$$

So the original integral is twice that, namely $\pi/5$.

9.

$$\lim_{a\to0^+}\int_a^1 \frac{x^4+1}{x}\,dx = \lim_{a\to0^+}(\frac{x^4}{4} + \ln x)\Big|_a^1 = \lim_{a\to0^+}[1/4 - (a^4/4 + \ln a)],$$

which diverges as $a \to 0$, since $\ln a \to -\infty$.

13.

$$\int_1^\infty \frac{y}{y^4+1}\,dy = \lim_{b\to\infty}\frac{1}{2}\int_1^b \frac{2y}{(y^2)^2+1}\,dy$$

$$= \lim_{b\to\infty}\frac{1}{2}\arctan(y^2)\Big|_1^b$$

$$= \lim_{b\to\infty}\frac{1}{2}[\arctan(b^2) - \arctan 1]$$

$$= (1/2)[\pi/2 - \pi/4] = \pi/8.$$

17.   As in Problem 14, $\int \dfrac{dx}{x \ln x} = \ln|\ln x| + C$, so

$$\int_1^2 \frac{dx}{x \ln x} = \lim_{b \to 1+} \int_b^2 \frac{dx}{x \ln x}$$

$$= \lim_{b \to 1+} \ln|\ln x| \Big|_b^2$$

$$= \lim_{b \to 1+} \ln(\ln 2) - \ln(\ln b).$$

As $b \to 1^+$, $\ln(\ln b) \to -\infty$, so the integral diverges.

21. $\displaystyle\int_4^\infty \frac{dx}{(x-1)^2} = \lim_{b \to \infty} \int_4^b \frac{dx}{(x-1)^2} = \lim_{b \to \infty} -\frac{1}{(x-1)} \Big|_4^b = \lim_{b \to \infty} \left[-\frac{1}{b-1} + \frac{1}{3} \right] = \frac{1}{3}.$

25.   Since the graph is above the x-axis for $x \geq 0$, we have

$$\text{Area} = \int_0^\infty xe^{-x}\, dx = \lim_{b \to \infty} \int_0^b xe^{-x}\, dx$$

$$= \lim_{b \to \infty} \left(-xe^{-x} \Big|_0^b + \int_0^b e^{-x}\, dx \right)$$

$$= \lim_{b \to \infty} \left(-be^{-b} - e^{-x} \Big|_0^b \right)$$

$$= \lim_{b \to \infty} (-be^{-b} - e^{-b} + e^0) = 1.$$

29.   (a)

$$\Gamma(1) = \int_0^\infty e^{-t}\, dt$$

$$= \lim_{b \to \infty} \int_0^b e^{-t}\, dt$$

$$= \lim_{b \to \infty} -e^{-t} \Big|_0^b$$

$$= \lim_{b \to \infty} [1 - e^{-b}] = 1.$$

Using Problem 2,

$$\Gamma(2) = \int_0^\infty te^{-t}\, dt = 1.$$

(b)   We integrate by parts. Let $u = t^n$, $v' = e^{-t}$. Then $u' = nt^{n-1}$ and $v = -e^{-t}$, so

$$\int t^n e^{-t}\, dt = -t^n e^{-t} + n \int t^{n-1} e^{-t}\, dt.$$

So

$$\Gamma(n+1) = \int_0^\infty t^n e^{-t}\, dt$$

$$= \lim_{b \to \infty} \int_0^b t^n e^{-t}\, dt$$

$$= \lim_{b \to \infty} \left[-t^n e^{-t} \Big|_0^b + n \int_0^b t^{n-1} e^{-t}\, dt \right]$$

$$= \lim_{b \to \infty} -b^n e^{-b} + \lim_{b \to \infty} n \int_0^b t^{n-1} e^{-t}\, dt$$

$$= 0 + n \int_0^\infty t^{n-1} e^{-t}\, dt$$

$$= n\Gamma(n).$$

(c) We already have $\Gamma(1) = 1$ and $\Gamma(2) = 1$. Using $\Gamma(n+1) = n\Gamma(n)$ we can get

$$\Gamma(3) = 2\Gamma(2) = 2$$
$$\Gamma(4) = 3\Gamma(3) = 3 \cdot 2$$
$$\Gamma(5) = 4\Gamma(4) = 4 \cdot 3 \cdot 2.$$

So it appears that $\Gamma(n)$ is just the first $n - 1$ numbers multiplied together, so $\Gamma(n) = (n-1)!$.

33. The factor $\ln x$ grows slowly enough (as $x \to 0^+$) not to change the convergence or divergence of the integral, although it will change what it converges or diverges to.

The integral is always improper, because $\ln x$ is not defined for $x = 0$. Integrating by parts (or, alternatively, the integral table) yields

$$\int_0^e x^p \ln x \, dx = \lim_{a \to 0^+} \int_a^e x^p \ln x \, dx$$

$$= \lim_{a \to 0^+} \left(\frac{1}{p+1} x^{p+1} \ln x - \frac{1}{(p+1)^2} x^{p+1} \right) \Bigg|_a^e$$

$$= \lim_{a \to 0^+} \left[\left(\frac{1}{p+1} e^{p+1} - \frac{1}{(p+1)^2} e^{p+1} \right) \right.$$

$$\left. - \left(\frac{1}{p+1} a^{p+1} \ln a - \frac{1}{(p+1)^2} a^{p+1} \right) \right].$$

If $p < -1$, then $(p+1)$ is negative, so as $a \to 0^+$, $a^{p+1} \to \infty$ and $\ln a \to -\infty$, and therefore the limit does not exist.

If $p > -1$, then $(p+1)$ is positive and it's easy to see that $a^{p+1} \to 0$ as $a \to 0$. Looking at graphs of $x^{p+1} \ln x$ (for different values of p) shows that $a^{p+1} \ln a \to 0$ as $a \to 0$. This isn't so easy to see analytically. It's true because if we let $t = \frac{1}{a}$ then

$$\lim_{a \to 0^+} a^{p+1} \ln a = \lim_{t \to \infty} \left(\frac{1}{t} \right)^{p+1} \ln \left(\frac{1}{t} \right) = \lim_{t \to \infty} -\frac{\ln t}{t^{p+1}}.$$

This last limit is zero because $\ln t$ grows very slowly, much more slowly than t^{p+1}. So if $p > -1$, the integral converges and equals $e^{p+1}[1/(p+1) - 1/(p+1)^2] = pe^{p+1}/(p+1)^2$.

What happens if $p = -1$? Then we get

$$\int_0^e \frac{\ln x}{x} \, dx = \lim_{a \to 0^+} \int_a^e \frac{\ln x}{x} \, dx$$

$$= \lim_{a \to 0^+} \frac{(\ln x)^2}{2} \Bigg|_a^e$$

$$= \lim_{a \to 0^+} \left(\frac{1 - (\ln a)^2}{2} \right).$$

Since $\ln a \to -\infty$ as $a \to 0^+$, this limit does not exist.

To summarize, $\int_0^e x^p \ln x$ converges for $p > -1$ to the value $pe^{p+1}/(p+1)^2$.

Solutions for Section 7.8

1. (a) The area is infinite. The area under $1/x$ is infinite and the area under $1/x^2$ is 1. So the area between the two has to be infinite also.

(b) Since $f(x)$ is bounded between 0 and $1/x^2$, and the area under $1/x^2$ is finite, $f(x)$ will have finite area by the comparison test. Similarly, $h(x)$ lies above $1/x$, whose area is infinite, so $h(x)$ must have infinite area as well. We can tell nothing about the area of $g(x)$, because the comparison test tells us nothing about a function larger than a function with finite area but smaller than one with infinite area. Finally, $k(x)$ will certainly have infinite area, because it has a lower bound m, for some $m > 0$. Thus, $\int_0^a k(x) \, dx \geq ma$, and since the latter does not converge as $a \to \infty$, neither can the former.

5. If $x \geq 1$, we know that $\dfrac{1}{x^3 + 1} \leq \dfrac{1}{x^3}$, and since $\displaystyle\int_1^\infty \dfrac{dx}{x^3}$ converges, the improper integral $\displaystyle\int_1^\infty \dfrac{dx}{x^3 + 1}$ converges.

9. This integral is convergent because, for $\phi \geq 1$,

$$\frac{2 + \cos\phi}{\phi^2} \leq \frac{3}{\phi^2},$$

and $\displaystyle\int_1^\infty \dfrac{3}{\phi^2}\,d\phi = 3\int_1^\infty \dfrac{1}{\phi^2}\,d\phi$ converges.

13. This integral is improper at $\theta = 0$. For $0 \leq \theta \leq 1$, we have $\dfrac{1}{\sqrt{\theta^3 + \theta}} \leq \dfrac{1}{\sqrt{\theta}}$, and since $\displaystyle\int_0^1 \dfrac{1}{\sqrt{\theta}}\,d\theta$ converges,

$\displaystyle\int_0^1 \dfrac{d\theta}{\sqrt{\theta^3 + \theta}}$ converges.

17. If we integrate e^{-x^2} from 1 to 10, we get 0.139. This answer doesn't change noticeably if you extend the region of integration to from 1 to 11, say, or even up to 1000. There's a reason for this; and the reason is that the tail, $\int_{10}^\infty e^{-x^2}\,dx$, is very small indeed. In fact

$$\int_{10}^\infty e^{-x^2}\,dx \leq \int_{10}^\infty e^{-x}\,dx = e^{-10},$$

which is very small. (In fact, the tail integral is less than $e^{-100}/10$. Can you prove that? [Hint: $e^{-x^2} \leq e^{-10x}$ for $x \geq 10$.])

21. (a) Since $e^{-x^2} \leq e^{-3x}$ for $x \geq 3$,

$$\int_3^\infty e^{-x^2}\,dx \leq \int_3^\infty e^{-3x}\,dx$$

Now

$$\int_3^\infty e^{-3x}\,dx = \lim_{b \to \infty} \int_3^b e^{-3x}\,dx = \lim_{b \to \infty} \left. -\frac{1}{3}e^{-3x} \right|_3^b$$

$$= \lim_{b \to \infty} \frac{e^{-9}}{3} - \frac{e^{-3b}}{3} = \frac{e^{-9}}{3}.$$

Thus

$$\int_3^\infty e^{-x^2}\,dx \leq \frac{e^{-9}}{3}.$$

(b) By reasoning similar to part (a),

$$\int_n^\infty e^{-x^2}\,dx \leq \int_n^\infty e^{-nx}\,dx,$$

and

$$\int_n^\infty e^{-nx}\,dx = \frac{1}{n}e^{-n^2},$$

so

$$\int_n^\infty e^{-x^2}\,dx \leq \frac{1}{n}e^{-n^2}.$$

25. (a) The tangent line to e^t has slope $(e^t)' = e^t$. Thus at $t = 0$, the slope is $e^0 = 1$. The line passes through $(0, e^0) = (0, 1)$. Thus the equation of the tangent line is $y = 1 + t$. Since e^t is everywhere concave up, its graph is always above the graph of any of its tangent lines; in particular, e^t is always above the line $y = 1 + t$. This is tantamount to saying

$$1 + t \leq e^t,$$

with equality holding only at the point of tangency, $t = 0$.

(b) If $t = \dfrac{1}{x}$, then the above inequality becomes

$$1 + \frac{1}{x} \le e^{1/x}, \text{ or } e^{1/x} - 1 \ge \frac{1}{x}.$$

Since $t = \dfrac{1}{x}$, t is never zero. Therefore, the inequality is strict, and we write

$$e^{1/x} - 1 > \frac{1}{x}.$$

(c) Since $e^{1/x} - 1 > \dfrac{1}{x}$,

$$\frac{1}{x^5\left(e^{1/x} - 1\right)} < \frac{1}{x^5\left(\frac{1}{x}\right)} = \frac{1}{x^4}.$$

Since $\displaystyle\int_1^\infty \frac{dx}{x^4}$ converges, $\displaystyle\int_1^\infty \frac{dx}{x^5\left(e^{1/x} - 1\right)}$ converges.

Solutions for Chapter 7 Review

1. The limits of integration are 0 and b, and the rectangle represents the region under the curve $f(x) = h$ between these limits. Thus,

$$\text{Area of rectangle} = \int_0^b h \, dx = hx \Big|_0^b = hb.$$

5. (a) i. 0   ii. $\frac{2}{\pi}$ iii. $\frac{1}{2}$
    (b) Average value of $f(t) <$ Average value of $k(t) <$ Average value of $g(t)$
        We can look at the three functions in the range $-\frac{\pi}{2} \le x \le \frac{3\pi}{2}$, since they all have periods of 2π ($|\cos t|$ and $(\cos t)^2$ also have a period of π, but that doesn't hurt our calculation). It is clear from the graphs of the three functions below that the average value for $\cos t$ is 0 (since the area above the x-axis is equal to the area below it), while the average values for the other two are positive (since they are everywhere positive, except where they are 0).

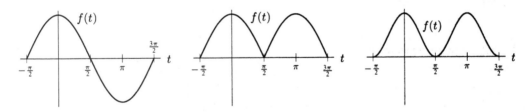

        It is also fairly clear from the graphs that the average value of $g(t)$ is greater than the average value of $k(t)$; it is also possible to see this algebraically, since

$$(\cos t)^2 = |\cos t|^2 \le |\cos t|$$

        because $|\cos t| \le 1$ (and both of these $\le$'s are $<$'s at all the points where the functions are not 0 or 1).

9. After the substitution $w = x + 2$, the first integral becomes

$$\int w^{-2} \, dw.$$

After the substitution $w = x^2 + 1$, the second integral becomes

$$\frac{1}{2} \int w^{-2} \, dw.$$

13. $\int_4^\infty \dfrac{dt}{t^{3/2}}$ should converge, since $\int_1^\infty \dfrac{dt}{t^n}$ converges for $n > 1$.
We calculate its value.

$$\int_4^\infty \frac{dt}{t^{3/2}} = \lim_{b \to \infty} \int_4^b t^{-3/2}\, dt = \lim_{b \to \infty} -2t^{-1/2}\Big|_4^b = \lim_{b \to \infty}\left(1 - \frac{2}{\sqrt{b}}\right) = 1.$$

17. Since the value of $\tan \theta$ is between -1 and 1 on the interval $-\pi/4 \le \theta \le \pi/4$, our integral is not improper and so converges. Moreover, since $\tan \theta$ is an odd function, we have

$$\int_{-\frac{\pi}{4}}^{\frac{\pi}{4}} \tan \theta\, d\theta = \int_{-\frac{\pi}{4}}^{0} \tan \theta\, d\theta + \int_{0}^{\frac{\pi}{4}} \tan \theta\, d\theta$$

$$= -\int_{-\frac{\pi}{4}}^{0} \tan(-\theta)\, d\theta + \int_{0}^{\frac{\pi}{4}} \tan \theta\, d\theta$$

$$= -\int_{0}^{\frac{\pi}{4}} \tan \theta\, d\theta + \int_{0}^{\frac{\pi}{4}} \tan \theta\, d\theta = 0.$$

21. Since $\sin \phi < \phi$ for $\phi > 0$,

$$\int_{0}^{\frac{\pi}{2}} \frac{1}{\sin \phi}\, d\phi > \int_{0}^{\frac{\pi}{2}} \frac{1}{\phi}\, d\phi,$$

The integral on the right diverges, so the integral on the left must also. Alternatively, we use IV-20 in the integral table to get

$$\int_{0}^{\frac{\pi}{2}} \frac{1}{\sin \phi}\, d\phi = \lim_{b \to 0^+} \int_{b}^{\frac{\pi}{2}} \frac{1}{\sin \phi}\, d\phi$$

$$= \lim_{b \to 0^+} \frac{1}{2} \ln \left| \frac{\cos \phi - 1}{\cos \phi + 1} \right| \Big|_{b}^{\frac{\pi}{2}}$$

$$= -\frac{1}{2} \lim_{b \to 0^+} \ln \left| \frac{\cos b - 1}{\cos b + 1} \right|.$$

As $b \to 0^+$, $\cos b - 1 \to 0$ and $\cos b + 1 \to 2$, so $\ln \left| \frac{\cos b - 1}{\cos b + 1} \right| \to -\infty$. Thus the integral diverges.

25. $\int_0^\pi \tan^2 \theta d\theta = \tan \theta - \theta + C$, by formula IV-23. The integrand blows up at $\theta = \frac{\pi}{2}$, so

$$\int_0^\pi \tan^2 \theta d\theta = \int_0^{\frac{\pi}{2}} \tan^2 \theta d\theta + \int_{\frac{\pi}{2}}^\pi \tan^2 \theta d\theta = \lim_{b \to \frac{\pi}{2}} [\tan \theta - \theta]_0^b + \lim_{a \to \frac{\pi}{2}} [\tan \theta - \theta]_a^\pi$$

which is undefined.

29.

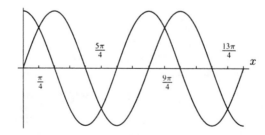

As is evident from the accompanying figure of the graphs of $y = \sin x$ and $y = \cos x$, the crossings occur at $x = \frac{\pi}{4}, \frac{5\pi}{4}, \frac{9\pi}{4}, \ldots$, and the regions bounded by any two consecutive crossings have the same area. So picking two consecutive crossings, we get an area of

$$\text{Area} = \int_{\frac{\pi}{4}}^{\frac{5\pi}{4}} (\sin x - \cos x)\, dx$$

$$= 2\sqrt{2}.$$

(Note that we integrated $\sin x - \cos x$ here because for $\frac{\pi}{4} \le x \le \frac{5\pi}{4}$, $\sin x \ge \cos x$.)

33. True. Since f' and g' are greater than 0, all left rectangles give underestimates. The bigger the derivative, the bigger the underestimate, so the bigger the error. (Note: if we didn't have $0 < f' < g'$, but instead just had $f' < g'$, the statement wouldn't necessarily be true. This is because some left rectangles could be overestimates and some could be underestimates–so, for example, it could be that the error in approximating g is 0! If $0 < f' < g'$, however, this can't happen.

37. Let us assume that SIMP(5) and (10) are both overestimates or both underestimates. Then since SIMP(10) is more accurate and bigger than SIMP(5), they are both underestimates. Since SIMP(10) is 16 times more accurate,

$$I - \text{SIMP}(5) = 16(I - \text{SIMP}(10)).$$

Solving for I, we have

$$I = \frac{16\text{SIMP}(10) - \text{SIMP}(5)}{15} \approx 7.4175.$$

Solutions to Practice Problems on Integration

1. Since $\dfrac{d}{dt}\cos t = -\sin t$, we have

$$\int \sin t \, dt = -\cos t + C, \quad \text{where } C \text{ is a constant.}$$

5. Since $\displaystyle\int \sin w \, d\theta = -\cos w + C$, the substitution $w = 2\theta$, $dw = 2\,d\theta$ gives $\displaystyle\int \sin 2\theta \, d\theta = -\frac{1}{2}\cos 2\theta + C.$

9. Either expand $(r+1)^3$ or use the substitution $w = r + 1$. If $w = r + 1$, then $dw = dr$ and

$$\int (r+1)^3 \, dr = \int w^3 \, dw = \frac{1}{4}w^4 + C = \frac{1}{4}(r+1)^4 + C.$$

13. Substitute $w = t^2$, so $dw = 2t\,dt$.

$$\int te^{t^2}\,dt = \frac{1}{2}\int e^{t^2}2t\,dt = \frac{1}{2}\int e^w\,dw = \frac{1}{2}e^w + C = \frac{1}{2}e^{t^2} + C.$$

Check:

$$\frac{d}{dt}\left(\frac{1}{2}e^{t^2} + C\right) = 2t\left(\frac{1}{2}e^{t^2}\right) = te^{t^2}.$$

17. Using substitution with $w = 1 - x$ and $dw = -dx$, we get

$$\int x\sqrt{1-x}\,dx = -\int (1-w)\sqrt{w}\,dw = \frac{2}{5}w^{5/2} - \frac{2}{3}w^{3/2} + C = \frac{2}{5}(1-x)^{5/2} - \frac{2}{3}(1-x)^{3/2} + C.$$

21. Remember that $\ln(x^2) = 2\ln x$. Therefore,

$$\int \ln(x^2)\,dx = 2\int \ln x\,dx = 2x\ln x - 2x + C.$$

Check:

$$\frac{d}{dx}(2x\ln x - 2x + C) = 2\ln x + \frac{2x}{x} - 2 = 2\ln x = \ln(x^2).$$

25. Expanding the numerator and dividing, we have

$$\int \frac{(u+1)^3}{u^2}\, du = \int \frac{(u^3 + 3u^2 + 3u + 1)}{u^2}\, du = \int \left(u + 3 + \frac{3}{u} + \frac{1}{u^2} \right) du$$

$$= \frac{u^2}{2} + 3u + 3\ln|u| - \frac{1}{u} + C.$$

Check:

$$\frac{d}{du}\left(\frac{u^2}{2} + 3u + 3\ln|u| - \frac{1}{u} + C \right) = u + 3 + 3/u + 1/u^2 = \frac{(u+1)^3}{u^2}.$$

29. Multiplying out and integrating term by term:

$$\int t^{10}(t-10)\, dt = \int (t^{11} - 10t^{10})\, dt = \int t^{11} dt - 10 \int t^{10}\, dt = \frac{1}{12}t^{12} - \frac{10}{11}t^{11} + C.$$

33. Integrating by parts, we take $u = e^{2x}$, $u' = 2e^{2x}$, $v' = \sin 2x$, and $v = -\frac{1}{2}\cos 2x$, so

$$\int e^{2x} \sin 2x\, dx = -\frac{e^{2x}}{2}\cos 2x + \int e^{2x}\cos 2x\, dx.$$

Integrating by parts again, with $u = e^{2x}$, $u' = 2e^{2x}$, $v' = \cos 2x$, and $v = \frac{1}{2}\sin 2x$, we get

$$\int e^{2x}\cos 2x\, dx = \frac{e^{2x}}{2}\sin 2x - \int e^{2x}\sin 2x\, dx.$$

Substituting into the previous equation, we obtain

$$\int e^{2x}\sin 2x\, dx = -\frac{e^{2x}}{2}\cos 2x + \frac{e^{2x}}{2}\sin 2x - \int e^{2x}\sin 2x\, dx.$$

Solving for $\int e^{2x}\sin 2x\, dx$ gives

$$\int e^{2x}\sin 2x\, dx = \frac{1}{4}e^{2x}(\sin 2x - \cos 2x) + C.$$

This result can also be obtained using II-8 in the integral table. Thus

$$\int_{-\pi}^{\pi} e^{2x}\sin 2x = [\frac{1}{4}e^{2x}(\sin 2x - \cos 2x)]\Big|_{-\pi}^{\pi} = \frac{1}{4}(e^{-2\pi} - e^{2\pi}) \approx -133.8724.$$

We get -133.37 using Simpson's rule with 10 intervals. With 100 intervals, we get -133.8724. Thus our answer matches the approximation of Simpson's rule.

37. Let $\sqrt{x} = w$, $\frac{1}{2}x^{-\frac{1}{2}}dx = dw$, $\frac{dx}{\sqrt{x}} = 2\, dw$. If $x = 1$ then $w = 1$, and if $x = 4$ so $w = 2$. So we have

$$\int_{1}^{4} \frac{e^{\sqrt{x}}}{\sqrt{x}}\, dx = \int_{1}^{2} e^w \cdot 2\, dw = 2e^w\Big|_{1}^{2} = 2(e^2 - e) \approx 9.34.$$

41. Integrating term by term:

$$\int \left(x^2 + 2x + \frac{1}{x} \right) dx = \frac{1}{3}x^3 + x^2 + \ln|x| + C,$$

where C is a constant.

45. If $u = \sin(5\theta)$, $du = \cos(5\theta) \cdot 5\,d\theta$, so

$$\int \sin(5\theta)\cos(5\theta)d\theta = \frac{1}{5}\int \sin(5\theta) \cdot 5\cos(5\theta)d\theta = \frac{1}{5}\int u\,du$$

$$= \frac{1}{5}\left(\frac{u^2}{2}\right) + C = \frac{1}{10}\sin^2(5\theta) + C$$

or

$$\int \sin(5\theta)\cos(5\theta)d\theta = \frac{1}{2}\int 2\sin(5\theta)\cos(5\theta)d\theta = \frac{1}{2}\int \sin(10\theta)d\theta \quad \text{(using } \sin(2x) = 2\sin x\cos x)$$

$$= \frac{-1}{20}\cos(10\theta) + C.$$

49. Let $w = \cos 2\theta$. Then $dw = -2\sin 2\theta\,d\theta$, hence

$$\int \cos^3 2\theta \sin 2\theta\,d\theta = -\frac{1}{2}\int w^3\,dw = -\frac{w^4}{8} + C = -\frac{\cos^4 2\theta}{8} + C.$$

Check:

$$\frac{d}{d\theta}\left(-\frac{\cos^4 2\theta}{8}\right) = -\frac{(4\cos^3 2\theta)(-\sin 2\theta)(2)}{8} = \cos^3 2\theta \sin 2\theta.$$

53. Let $\sin\theta = w$, then $\cos\theta\,d\theta = dw$, so

$$\int \cos\theta\sqrt{1 + \sin\theta}\,d\theta = \int \sqrt{1 + w}\,dw$$

$$= \frac{(1 + w)^{3/2}}{3/2} + C = \frac{2}{3}(1 + \sin\theta)^{3/2} + C,$$

where C is a constant.

57. Let $w = 3z + 5$ and $dw = 3\,dz$. Then

$$\int (3z + 5)^3\,dz = \frac{1}{3}\int w^3\,dw = \frac{1}{12}w^4 + C = \frac{1}{12}(3z + 5)^4 + C.$$

61. Let $w = \ln x$, then $dw = (1/x)dx$ so that

$$\int \frac{1}{x}\sin(\ln x)\,dx = \int \sin w\,dw = -\cos w + C = -\cos(\ln x) + C.$$

65. Let $u = 1 - \cos w$, then $du = \sin w\,dw$ which gives

$$\int \frac{\sin w\,dw}{\sqrt{1 - \cos w}} = \int \frac{du}{\sqrt{u}} = 2\sqrt{u} + C = 2\sqrt{1 - \cos w} + C.$$

69. Integrating by parts using $u = t^2$ and $dv = \frac{t\,dt}{\sqrt{1+t^2}}$ gives $du = 2t\,dt$ and $v = \sqrt{1 + t^2}$. Now

$$\int \frac{t^3}{\sqrt{1 + t^2}}\,dt = t^2\sqrt{1 + t^2} - \int 2t\sqrt{1 + t^2}\,dt$$

$$= t^2\sqrt{1 + t^2} - \frac{2}{3}(1 + t^2)^{3/2} + C$$

$$= \sqrt{1 + t^2}(t^2 - \frac{2}{3}(1 + t^2)) + C$$

$$= \sqrt{1 + t^2}\frac{(t^2 - 2)}{3} + C.$$

73.   Integrate by parts letting $u = (\ln r)^2$ and $dv = r\,dr$, then $du = (2/r)\ln r\,dr$ and $v = r^2/2$. We get

$$\int r(\ln r)^2\,dr = \frac{1}{2}r^2(\ln r)^2 - \int r\ln r\,dr.$$

Then using integration by parts again with $u = \ln r$ and $dv = r\,dr$, so $du = dr/r$ and $v = r^2/2$, we get

$$\int r\ln^2 r\,dr = \frac{1}{2}r^2(\ln r)^2 - \left[\frac{1}{2}r^2\ln r - \frac{1}{2}\int r\,dr\right] = \frac{1}{2}r^2(\ln r)^2 - \frac{1}{2}r^2\ln r + \frac{1}{4}r^2 + C.$$

77.   Using Table IV-19, let $m = 3$, $w = 2x$, and $dw = 2\,dx$. Then

$$\int \frac{1}{\sin^3(2x)}\,dx = \frac{1}{2}\int \frac{1}{\sin^3 w}\,dw$$

$$= \frac{1}{2}\left[\frac{-1}{(3-1)}\frac{\cos w}{\sin^2 w}\right] + \frac{1}{4}\int \frac{1}{\sin w}\,dw,$$

and using Table IV-20, we have

$$\int \frac{1}{\sin w}\,dw = \frac{1}{2}\ln\left|\frac{\cos w - 1}{\cos w + 1}\right| + C.$$

Thus,

$$\int \frac{1}{\sin^3(2x)}\,dx = -\frac{\cos 2x}{4\sin^2 2x} + \frac{1}{8}\ln\left|\frac{\cos 2x - 1}{\cos 2x + 1}\right| + C.$$

81.   Using II-9 from the integral table, with $a = 5$ and $b = 3$, we have

$$\int e^{5x}\cos(3x)\,dx = \frac{1}{25+9}e^{5x}\left[5\cos(3x) + 3\sin(3x)\right] + C$$

$$= \frac{1}{34}e^{5x}\left[5\cos(3x) + 3\sin(3x)\right] + C.$$

85.   We know $x^2 + 5x + 4 = (x+1)(x+4)$, so we can use V-26 of the integral table with $a = -1$ and $b = -4$ to write

$$\int \frac{dx}{x^2 + 5x + 4} = \frac{1}{3}(\ln|x+1| - \ln|x+4|) + C.$$

89.   We can factor the denominator into $ax(x + \frac{b}{a})$, so

$$\int \frac{dx}{ax^2 + bx} = \frac{1}{a}\int \frac{1}{x(x + \frac{b}{a})}$$

Now we can use V-26 (with $A = 0$ and $B = -\frac{b}{a}$ to give

$$\frac{1}{a}\int \frac{1}{x(x + \frac{b}{a})} = \frac{1}{a}\cdot\frac{a}{b}\left(\ln|x| - \ln\left|x + \frac{b}{a}\right|\right) + C = \frac{1}{b}\left(\ln|x| - \ln\left|x + \frac{b}{a}\right|\right) + C.$$

93.   If $u = 2^t + 1$, $du = 2^t(\ln 2)\,dt$, so

$$\int \frac{2^t}{2^t + 1}\,dt = \frac{1}{\ln 2}\int \frac{2^t\ln 2}{2^t + 1}\,dt = \frac{1}{\ln 2}\int \frac{1}{u} = \frac{1}{\ln 2}\ln|u| + C = \frac{1}{\ln 2}\ln|2^t + 1| + C.$$

97.   Let $x = 2\theta$, then $dx = 2d\theta$. Thus

$$\int \sin^2(2\theta) \cos^3(2\theta) \, d\theta = \frac{1}{2} \int \sin^2 x \cos^3 x \, dx.$$

We let $w = \sin x$ and $dw = \cos x \, dx$. Then

$$\frac{1}{2} \int \sin^2 x \cos^3 x \, dx = \frac{1}{2} \int \sin^2 x \cos^2 x \cos x \, dx$$

$$= \frac{1}{2} \int \sin^2 x (1 - \sin^2 x) \cos x \, dx$$

$$= \frac{1}{2} \int w^2(1 - w^2) \, dw = \frac{1}{2} \int (w^2 - w^4) \, dw$$

$$= \frac{1}{2} \left(\frac{w^3}{3} - \frac{w^5}{5} \right) + C = \frac{1}{6} \sin^3 x - \frac{1}{10} \sin^5 x + C$$

$$= \frac{1}{6} \sin^3(2\theta) - \frac{1}{10} \sin^5(2\theta) + C.$$

101.   If $u = 1 + \cos^2 w$, $du = 2(\cos w)^1(-\sin w) \, dw$, so

$$\int \frac{\sin w \cos w}{1 + \cos^2 w} \, dw = -\frac{1}{2} \int \frac{-2 \sin w \cos w}{1 + \cos^2 w} \, dw = -\frac{1}{2} \int \frac{1}{u} \, du = -\frac{1}{2} \ln |u| + C$$

$$= -\frac{1}{2} \ln |1 + \cos^2 w| + C.$$

105.   If $u = \sqrt{x + 1}$, $u^2 = x + 1$ with $x = u^2 - 1$ and $dx = 2u \, du$. Substituting, we get

$$\int \frac{x}{\sqrt{x + 1}} \, dx = \int \frac{(u^2 - 1)2u \, du}{u} = \int (u^2 - 1)2 \, du = 2 \int (u^2 - 1) \, du$$

$$= \frac{2u^3}{3} - 2u + C = \frac{2(\sqrt{x + 1})^3}{3} - 2\sqrt{x + 1} + C.$$

109.   Letting $u = z - 5$, $z = u + 5$, $dz = du$, and substituting, we have

$$\int \frac{z}{(z - 5)^3} dz = \int \frac{u + 5}{u^3} du = \int (u^{-2} + 5u^{-3}) du = \frac{u^{-1}}{-1} + 5 \left(\frac{u^{-2}}{-2} \right) + C$$

$$= \frac{-1}{(z - 5)} + \frac{-5}{2(z - 5)^2} + C.$$

113.   Using Table III-16,

$$\int (2x^3 + 3x + 4) \cos(2x) \, dx = \frac{1}{2}(2x^3 + 3x + 4) \sin(2x)$$

$$+ \frac{1}{4}(6x^2 + 3) \cos(2x)$$

$$- \frac{1}{8}(12x) \sin(2x) - \frac{3}{4} \cos(2x) + C.$$

$$= 2 \sin(2x) + x^3 \sin(2x) + \frac{3x^2}{2} \cos(2x) + C.$$

CHAPTER EIGHT

Solutions for Section 8.1

1. Vertical slices are circular. Horizontal slices would be similar to ellipses in cross-section, or at least ovals (a word derived from *ovum*, the Latin word for egg).

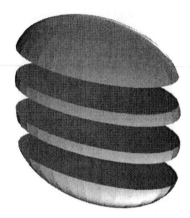

Figure 8.1

5.

Radius $= \frac{b\sqrt{a^2-x^2}}{a}$

$$y^2 = b^2\left(1 - \frac{x^2}{a^2}\right).$$

$$V = \int_{-a}^{a} \pi y^2 \, dx = \pi \int_{-a}^{a} b^2 \left(1 - \frac{x^2}{a^2}\right) dx$$

$$= 2\pi b^2 \int_{0}^{a}\left(1 - \frac{x^2}{a^2}\right) dx = 2\pi b^2 \left[x - \frac{x^3}{3a^2}\right]_0^a$$

$$= 2\pi b^2 \left(a - \frac{a^3}{3a^2}\right) = 2\pi b^2 \left(a - \frac{1}{3}a\right)$$

$$= \frac{4}{3}\pi ab^2.$$

9.

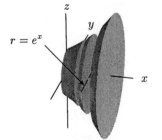

$r = e^x$

This is the volume of revolution gotten from the rotating the curve $y = e^x$. Take slices perpendicular to the x-axis. They will be circles with radius e^x, so

$$V = \int_{x=0}^{x=1} \pi y^2 \, dx = \pi \int_0^1 e^{2x} \, dx$$

$$= \left.\frac{\pi e^{2x}}{2}\right|_0^1 = \frac{\pi(e^2 - 1)}{2} \approx 10.036.$$

13.

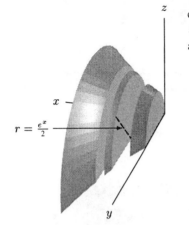

We slice perpendicular to the x-axis. As stated in the problem, the cross-sections obtained thereby will be semicircles, with radius $\frac{e^x}{2}$. The volume of one semicircular slice is $\frac{1}{2}\pi\left(\frac{e^x}{2}\right)^2 dx$. (Look at the picture.) Adding up the volumes of the slices yields

$$\text{Volume} = \int_{x=0}^{x=1} \pi\frac{y^2}{2}\,dx = \frac{\pi}{8}\int_0^1 e^{2x}\,dx$$

$$= \left.\frac{\pi e^{2x}}{16}\right|_0^1 = \frac{\pi(e^2-1)}{16} \approx 1.25.$$

17.   We can find the volume of the tree by slicing it into a series of thin horizontal cylinders of height dh and circumference C. The volume of each cylindrical disk will then be

$$V = \pi r^2\,dh = \pi\left(\frac{C}{2\pi}\right)^2 dh = \frac{C^2\,dh}{4\pi}.$$

Summing all such cylinders, we have the total volume of the tree as

$$\text{Total volume} = \frac{1}{4\pi}\int_0^{120} C^2\,dh.$$

We can estimate this volume using a trapezoidal approximation to the integral with $\Delta h = 20$:

$$\text{LEFT estimate} = \frac{1}{4\pi}[20(31^2 + 28^2 + 21^2 + 17^2 + 12^2 + 8^2)] = \frac{1}{4\pi}(53660).$$

$$\text{RIGHT estimate} = \frac{1}{4\pi}[20(28^2 + 21^2 + 17^2 + 12^2 + 8^2 + 2^2)] = \frac{1}{4\pi}(34520).$$

$$\text{TRAP} = \frac{1}{4\pi}(44090) \approx 3509 \text{ cubic inches.}$$

21.   (a)   The equation of a circle of radius r around the origin is $x^2 + y^2 = r^2$. This means that $y^2 = r^2 - x^2$, so $2y(dy/dx) = -2x$, and $dy/dx = -x/y$. Since the circle is symmetric about both axes, its arc length is 4 times the arc length in the first quadrant, namely

$$4\int_0^r \sqrt{1 + \left(\frac{dy}{dx}\right)^2}\,dx = 4\int_0^r \sqrt{1 + \left(-\frac{x}{y}\right)^2}\,dx.$$

(b)   Evaluating this integral yields

$$4\int_0^r \sqrt{1 + \left(-\frac{x}{y}\right)^2}\,dx = 4\int_0^r \sqrt{1 + \frac{x^2}{r^2 - x^2}}\,dx = 4\int_0^r \sqrt{\frac{r^2}{r^2 - x^2}}\,dx$$

$$= 4r\int_0^r \sqrt{\frac{1}{r^2 - x^2}}\,dx = \left.4r(\arcsin(x/r))\right|_0^r = 2\pi r.$$

This is the expected answer.

25.   Here are many functions which "work."

- Any linear function $y = mx + b$ "works." This follows because $\frac{dy}{dx} = m$ is constant for such functions. So

$$\int_a^b \sqrt{1 + \left(\frac{dy}{dx}\right)^2}\,dx = \int_a^b \sqrt{1 + m^2}\,dx = (b - a)\sqrt{1 + m^2}.$$

- The function $y = \frac{x^4}{8} + \frac{1}{4x^2}$ "works": $\frac{dy}{dx} = \frac{1}{2}(x^3 - 1/x^3)$, and

$$\int \sqrt{1 + \left(\frac{dy}{dx}\right)^2}\, dx = \int \sqrt{1 + \frac{\left(x^3 - \frac{1}{x^3}\right)^2}{4}}\, dx = \int \sqrt{1 + \frac{x^6}{4} - \frac{1}{2} + \frac{1}{4x^6}}\, dx$$

$$= \int \sqrt{\frac{1}{4}\left(x^3 + \frac{1}{x^3}\right)^2}\, dx = \int \frac{1}{2}\left(x^3 + \frac{1}{x^3}\right)\, dx$$

$$= \left[\frac{x^4}{8} - \frac{1}{4x^2}\right] + C.$$

- One more function that "works" is $y = \ln(\cos x)$; we have $\frac{dy}{dx} = -\sin x / \cos x$. Hence

$$\int \sqrt{1 + \left(\frac{dy}{dx}\right)^2}\, dx. = \int \sqrt{1 + \left(\frac{-\sin x}{\cos x}\right)^2}\, dx = \int \sqrt{1 + \frac{\sin^2 x}{\cos^2 x}}\, dx$$

$$= \int \sqrt{\frac{\sin^2 x + \cos^2 x}{\cos^2 x}}\, dx = \int \sqrt{\frac{1}{\cos^2 x}}\, dx$$

$$= \int \frac{1}{\cos x}\, dx = \frac{1}{2}\ln\left|\frac{\sin x + 1}{\sin x - 1}\right| + C,$$

where the last integral comes from IV-22 of the integral tables.

Solutions for Section 8.2

1. (a) Suppose we choose an x, $0 \le x \le 2$. If Δx is a small fraction of a meter, then the density of the rod is approximately $\rho(x)$ anywhere from x to $x + \Delta x$ meters from the left end of the rod (see below). The mass of the rod from x to $x + \Delta x$ meters is therefore approximately $\rho(x)\Delta x = (2 + 6x)\Delta x$. If we slice the rod into N pieces, then a Riemann sum is $\sum_{i=1}^{N} (2 + 6xi)\Delta x$.

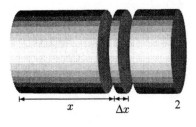

(b) The definite integral is

$$M = \int_0^2 \rho(x)\, dx = \int_0^2 (2 + 6x)\, dx = (2x + 3x^2)\Big|_0^2 = 16 \text{ grams.}$$

5. (a) We form a Riemann sum by slicing the region into concentric rings of radius r and width Δr. Then the volume deposited on one ring will be the height $H(r)$ multiplied by the area of the ring. A ring of width Δr will have an area given by

$$\text{Area} = \pi(r + \Delta r)^2 - \pi(r^2)$$
$$= \pi(r^2 + 2r\Delta r + (\Delta r)^2 - r^2)$$
$$= \pi(2r\Delta r + (\Delta r)^2).$$

Since Δr is approaching zero, we can approximate

$$\text{Area of ring} \approx \pi(2r\Delta r + 0) = 2\pi r\Delta r.$$

From this, we have

$$\Delta V \approx H(r) \cdot 2\pi r\Delta r.$$

Thus, summing the contributions from all rings we have

$$V \approx \sum H(r) \cdot 2\pi r\Delta r.$$

Taking the limit as $\Delta r \to 0$, we get

$$V = \int_0^5 2\pi r \left(0.115e^{-2r}\right) dr.$$

(b) We use integration by parts:

$$V = 0.23\pi \int_0^5 \left(re^{-2r}\right) dr$$

$$= 0.23\pi \left(\frac{re^{-2r}}{-2} - \frac{e^{-2r}}{4}\right)\Bigg|_0^5$$

$$\approx 0.181(\text{millimeters}) \cdot (\text{kilometers})^2 = 0.181 \cdot 10^{-3} \cdot 10^6 \text{ meters}^3 = 181 \text{ cubic meters}.$$

9. (a) The density is minimum at $x = -1$ and increases as x increases, so more of the mass of the rod is in the right half of the rod. We thus expect the balancing point to be to the right of the origin.

(b) We need to compute

$$\int_{-1}^1 x(3 - e^{-x})\, dx = \left(\frac{3}{2}x^2 + xe^{-x} + e^{-x}\right)\Bigg|_{-1}^1 \quad \text{(using integration by parts)}$$

$$= \frac{3}{2} + e^{-1} + e^{-1} - \left(\frac{3}{2} - e^1 + e^1\right) = \frac{2}{e}.$$

We must divide this result by the total mass, which is given by

$$\int_{-1}^1 (3 - e^{-x})\, dx = (3x + e^{-x})\Bigg|_{-1}^1 = 6 - e + \frac{1}{e}.$$

We therefore have

$$\bar{x} = \frac{2/e}{6 - e + (1/e)} = \frac{2}{1 + 6e - e^2} \approx 0.2.$$

13. (a) Use the formula for the volume of a cylinder:

$$\text{Volume} = \pi r^2 l.$$

Since it is only a half cylinder

$$\text{Volume of shed} = \frac{1}{2}\pi r^2 l.$$

(b) Set up the axes as shown in Figure 8.2. The density can be defined as

$$\text{Density} = ky.$$

Now slice the sawdust horizontally into slabs of thickness Δy as shown in Figure 8.3, and calculate the volume of each slab:

$$\text{Volume of slab} \approx 2xl\Delta y = 2l(\sqrt{r^2 - y^2})\Delta y.$$

Finally, we compute the total mass of sawdust:

$$\text{Total mass of sawdust} = \int_0^r 2kly\sqrt{r^2 - y^2}\, dy = -\frac{2}{3}kl(r^2 - y^2)^{3/2}\Bigg|_0^r = \frac{2klr^3}{3}.$$

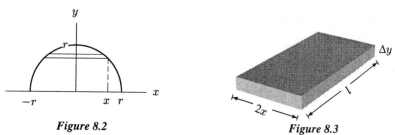

Figure 8.2   **Figure 8.3**

Solutions for Section 8.3

1. (a) Looking at the graph, it appears that the graph of B is above $F = 10$ between $t = 2.3$ and $t = 4.2$, or for about 1.9 seconds.
    (b) Although the total impulse is defined as the integral from 0 to ∞, the thrust is 0 after a certain time, so the integral is actually not improper. From $t = 0$ to $t = 2$, the graph of A looks like a triangle with base 2 and height 12, for an area of 12. From $t = 2$ to $t = 4$, the graph of A looks a trapezoid with base 2 and heights 13 and 6, for an area of 19. From $t = 4$ to $t = 16$, A is approximately a rectangle with height 5.8 and width 12, for an area of 69.6. Finally, from $t = 16$ to $t = 17$, A looks like a triangle with base 1 and height 5.8, for an area of 2.9. So, the total area under the curve of A's thrust, which is A's total impulse, is about 103.5 newton-seconds.
    (c) Note that when we calculated the impulse in part (b), we multiplied height, measured in newtons, by width, measured in seconds. So the units of impulse are newton-seconds.
    (d) The graph of B's thrust looks like a triangle with base 6 and height 22, for a total impulse of about 66 newton-seconds. So rocket A, with total impulse 103.5 newton-seconds, has a larger total impulse than rocket B.
    (e) As we can see from the graph, rocket B reaches a maximum thrust of 22, whereas A only reaches a maximum thrust of 13. So rocket B has the largest maximum thrust.

5. Let x be the distance measured from the bottom the tank. To pump a layer of water of thickness Δx at x feet from the bottom, the work needed is

    $$(62.4)\pi 6^2 (20 - x)\Delta x.$$

    Therefore, the total work is

    $$W = \int_0^{10} 36 \cdot (62.4)\pi(20 - x)dx$$
    $$= 36 \cdot (62.4)\pi\left(20x - \frac{1}{2}x^2\right)\Big|_0^{10}$$
    $$= 36 \cdot (62.4)\pi(200 - 50)$$
    $$\approx 1,058,591.1 \text{ ft-lb.}$$

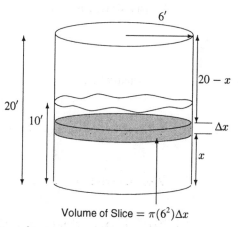

Volume of Slice $= \pi(6^2)\Delta x$

9. The force exerted on the satellite by the earth (and vice versa!) is GMm/r^2, where r is the distance from the center of the earth to the center of the satellite, m is the mass of the satellite, M is the mass of the earth, and G is the gravitational constant. So the total work done is

    $$\int_{6.4\times 10^6}^{8.4\times 10^6} \frac{GMm}{r^2}\, dr = \left(\frac{-GMm}{r}\right)\Big|_{6.4\times 10^6}^{8.4\times 10^6} \approx 1.489 \times 10^{10} \text{ joules.}$$

13.

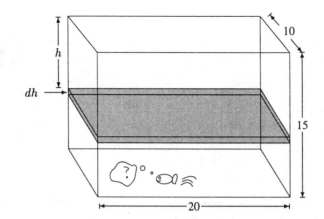

- Bottom: The bottom of the tank is at constant depth 15 feet, and therefore is under constant pressure, $15 \cdot 62.4 =$ 936 lb/ft^2. The area of the base is 200 ft^2 and so the total force is 200 ft$^2 \cdot$ 936 lb/ft^2 = 187200 lb.
- 15×10 side: The area of a horizontal strip of width dh is $10\,dh$ square feet, and the pressure at height h is 62.4h pounds per square foot. Therefore, the force on such a strip is 62.4$h(10\,dh)$ pounds. Hence, the total force on this side is

$$\int_0^{15} (62.4h)(10)\,dh = 624\frac{h^2}{2}\bigg|_0^{15} = 70200 \text{ lbs.}$$

- 15×20 side: Similarly, the total force on this side is

$$\int_0^{15} (62.4h)(20)\,dh = 1248\frac{h^2}{2}\bigg|_0^{15} = 140400 \text{ lbs.}$$

17. We need to divide the disk up into circular rings of charge and integrate their contributions to the potential (at P) from 0 to a. These rings, however, are not uniformly distant from the point P. A ring of radius z is $\sqrt{R^2 + z^2}$ away from point P (see below).

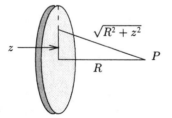

The ring has area $2\pi z\,\Delta z$, and charge $2\pi z \sigma\,\Delta z$. The potential of the ring is then $\dfrac{2\pi z\sigma\,\Delta z}{\sqrt{R^2 + z^2}}$ and the total potential at point P is

$$\int_0^a \frac{2\pi z\sigma\,dz}{\sqrt{R^2 + z^2}} = \pi\sigma \int_0^a \frac{2z\,dz}{\sqrt{R^2 + z^2}}.$$

We make the substitution $u = z^2$. Then $du = 2z\,dz$. We obtain

$$\pi\sigma \int_0^a \frac{2z\,dz}{\sqrt{R^2 + z^2}} = \pi\sigma \int_0^{a^2} \frac{du}{\sqrt{R^2 + u}} = \pi\sigma(2\sqrt{R^2 + u})\bigg|_0^{a^2}$$

$$= \pi\sigma(2\sqrt{R^2 + z^2})\bigg|_0^a = 2\pi\sigma(\sqrt{R^2 + a^2} - R).$$

(The substitution $u = R^2 + z^2$ or $\sqrt{R^2 + z^2}$ works also.)

21.

This time, let's split the second rod into small slices of length dr. Each slice is of mass $\frac{M_2}{l_2} dr$, since the density of the second rod is $\frac{M_2}{l_2}$. Since the slice is small, we can treat it as a particle at distance r away from the end of the first rod, as in Problem 20. By that problem, the force of attraction between the first rod and particle is

$$\frac{GM_1 \frac{M_2}{l_2} dr}{(r)(r + l_1)}.$$

So the total force of attraction between the rods is

$$\int_a^{a+l_2} \frac{GM_1 \frac{M_2}{l_2} dr}{(r)(r + l_1)} = \frac{GM_1 M_2}{l_2} \int_a^{a+l_2} \frac{dr}{(r)(r + l_1)}$$

$$= \frac{GM_1 M_2}{l_2} \int_a^{a+l_2} \frac{1}{l_1} \left(\frac{1}{r} - \frac{1}{r + l_1} \right) dr.$$

$$= \frac{GM_1 M_2}{l_1 l_2} \left(\ln|r| - \ln|r + l_1| \right) \Big|_a^{a+l_2}$$

$$= \frac{GM_1 M_2}{l_1 l_2} \left[\ln|a + l_2| - \ln|a + l_1 + l_2| - \ln|a| + \ln|a + l_1| \right]$$

$$= \frac{GM_1 M_2}{l_1 l_2} \ln \left[\frac{(a + l_1)(a + l_2)}{a(a + l_1 + l_2)} \right].$$

This result is symmetric: if you switch l_1 and l_2 or M_1 and M_2, you get the same answer. That means it's not important which rod is "first," and which is "second."

Solutions for Section 8.4

1.

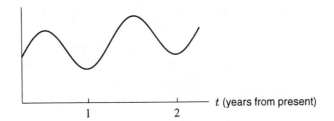

The graph reaches a peak each summer, and a trough each winter. The graph shows sunscreen sales increasing from cycle to cycle. This gradual increase may be due in part to inflation and to population growth.

5. (a) Solve for $P(t) = P$.

$$100000 = \int_0^{10} Pe^{0.10(10-t)} dt = Pe \int_0^{10} e^{-0.10t} dt$$

$$= \frac{Pe}{-0.10} e^{-0.10t} \Big|_0^{10} = Pe(-3.678 + 10)$$

$$= P \cdot 17.183.$$

So, $P \approx \$5820$ per year.

(b) To answer this, we'll calculate the present value of $100,000:

$$100000 = Pe^{0.10(10)}$$

$$P \approx \$36,787.94.$$

9. (a)

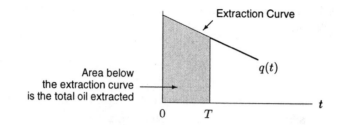

Suppose the oil extracted over the time period $[0, T]$ is S. (See above.)Since $q(t)$ is the rate of oil extraction, we have:

$$S = \int_0^T q(t)dt = \int_0^T (a - bt)dt = \left(at - \frac{bt^2}{2} \right) \Big|_0^T = aT - \frac{b}{2}T^2.$$

We know $a = 10, b = 0.1$. To calculate the time at which the oil is exhausted, we set $S = 100$ and we solve for T:

$$T = \frac{a \pm \sqrt{a^2 - 2bS}}{b} = 10.6 \quad \text{or} \quad 189.4.$$

Since $T = 189.4$ makes $q(t) < 0$, the answer is $T = 10.6$ years.

(b) Suppose p is the oil price, C is the extraction cost per barrel, and r is the interest rate. We have the present value of the profit as

$$\text{Present value of profit} = \int_0^T (p - C)q(t)e^{-rt}dt$$

$$= (p - C) \int_0^T (a - bt)e^{-rt}dt$$

$$= \frac{p - C}{r^2}[ar - b - (ar - brT - b)e^{-rT}].$$

Since $p = 20, C = 10$, and $r = 0.1$, we have

$$\text{Present value of profit} = 624.9 \text{ million dollars.}$$

13.

$$\int_0^{q^*} (p^* - S(q)) \, dq = \int_0^{q^*} p^* \, dq - \int_0^{q^*} S(q) \, dq$$

$$= p^* q^* - \int_0^{q^*} S(q) \, dq.$$

Using Problem 12, this integral is the extra amount consumers pay (i.e., suppliers earn over and above the minimum they would be willing to accept for supplying the good). It results from charging the equilibrium price.

Solutions for Chapter 8 Review

1. (a)

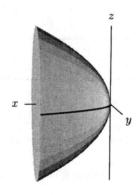

(b)   Divide $[0,1]$ into N subintervals of width $\Delta x = \frac{1}{N}$. The volume of the ith disc is $\pi(\sqrt{x_i})^2 \Delta x = \pi x_i \Delta x$. So, $V \approx \sum_{i=1}^{N} \pi x_i \Delta x$.

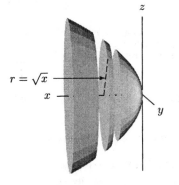

(c)

$$\text{Volume} = \int_0^1 \pi x \, dx = \left.\frac{\pi}{2} x^2\right|_0^1 = \frac{\pi}{2} \approx 1.57.$$

5.   (a)   If you slice the apple perpendicular to the core, you expect that the cross section will be approximately a circle.

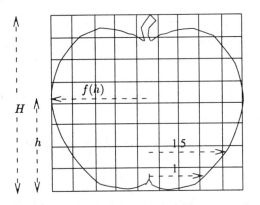

If $f(h)$ is the radius of the apple at height h above the bottom, and H is the height of the apple, then

$$\text{Volume} = \int_0^H \pi f(h)^2 \, dh.$$

Ignoring the stem, $H \approx 3.5$. Although we do not have a formula for $f(h)$, we can estimate it at various points. (Remember, we measure here from the bottom of the *apple*, which is not quite the bottom of the graph.)

h	0	0.5	1	1.5	2	2.5	3	3.5
$f(h)$	1	1.5	2	2.1	2.3	2.2	1.8	1.2

Now let $g(h) = \pi f(h)^2$, the area of the cross-section at height h. From our approximations above, we get the following table.

h	0	0.5	1	1.5	2	2.5	3	3.5
$g(h)$	3.14	7.07	12.57	13.85	16.62	13.85	10.18	4.52

We can now take left- and right-hand sum approximations. Note that $\Delta h = 0.5$ inches. Thus

$$\text{LEFT}(9) = (3.14 + 7.07 + 12.57 + 13.85 + 16.62 + 13.85 + 10.18)(0.5) = 38.64.$$
$$\text{RIGHT}(9) = (7.07 + 12.57 + 13.85 + 16.62 + 13.85 + 10.18 + 4.52)(0.5) = 39.33.$$

Thus the volume of the apple is ≈ 39 cu.in.

(b)   The apple weighs $0.03 \times 39 \approx 1.17$ pounds, so it costs about 94¢. (Expensive apple!)

9. We'll divide up time between 1971 and 1992 into intervals of length dt, and figure out how much of the strontium-90 produced during that time interval is still around.

    First, strontium-90 decays exponentially, so if a quantity S_0 was produced t years ago, and S is the quantity around today, $S = S_0 e^{-kt}$. Since the half-life is 28 years, $\frac{1}{2} = e^{-k(28)}$, giving $k = \frac{-\ln(\frac{1}{2})}{28} \approx 0.025$.

    Suppose we measure t in years from 1971, so that 1992 is $t = 21$.

    Since strontium-90 is produced at a rate of 1 kg/year, during the interval dt we know that a quantity $1dt$ kg was produced. Since this was $(21 - t)$ years ago, the quantity remaining now is $1dt e^{-0.025(21-t)}$. Summing over all such intervals gives

    $$\text{Strontium in 1992} \approx \int_0^{21} e^{-0.025(21-t)}\, dt$$
    $$= \left. \frac{e^{-0.025(21-t)}}{0.025} \right|_0^{21} = 16.34 \text{ kg.}$$

    [Note: This is exactly like a future value problem from economics, with a negative interest rate.]

13.

    $$\text{Water force} = \int_0^{25} (62.4)(25 - h)(60)\, dh$$
    $$= (62.4)(60) \left. \left(25h - \frac{h^2}{2} \right) \right|_0^{25}$$
    $$= (62.4)(60)(625 - 312.5)$$
    $$= (62.4)(60)(312.5)$$
    $$= 1,170,000 \text{ lbs}$$

17. (a)   Slice the mountain horizontally into N cylinders of height Δh. The sum of the volumes of the cylinders will be

    $$\sum_{i=1}^{N} \pi r^2 \Delta h = \sum_{i=1}^{N} \pi \left(\frac{3.5 \cdot 10^5}{\sqrt{h + 600}} \right)^2 \Delta h.$$

    (b)

    $$\text{Volume} = \int_{400}^{14400} \pi \left(\frac{3.5 \cdot 10^5}{\sqrt{h + 600}} \right)^2 dh$$
    $$= 1.23 \cdot 10^{11} \pi \int_{400}^{14400} \frac{1}{(h + 600)}\, dh$$
    $$= 1.23 \cdot 10^{11} \pi \left. \ln(h + 600) \right|_{400}^{14400} dh$$
    $$= 1.23 \cdot 10^{11} \pi \left[\ln 15000 - \ln 1000 \right]$$
    $$= 1.23 \cdot 10^{11} \pi \ln(15000/1000)$$
    $$= 1.23 \cdot 10^{11} \pi \ln 15 \approx 1.05 \cdot 10^{12} \text{ cubic feet.}$$

21. (a) Slicing horizontally, as shown in Figure 8.4, we see that the volume of one disk-shaped slab is

$$\Delta V \approx \pi x^2 \Delta y = \frac{\pi y}{a} \Delta y.$$

Thus, the volume of the water is given by

$$V = \int_0^h \frac{\pi}{a} y \, dy = \frac{\pi}{a} \frac{y^2}{2} \Big|_0^h = \frac{\pi h^2}{2a}.$$

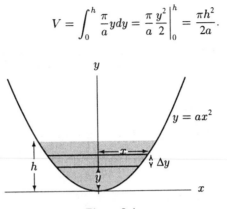

Figure 8.4

(b) The surface of the water is a circle of radius x. Since at the surface, $y = h$, we have $h = ax^2$. Thus, at the surface, $x = \sqrt{(h/a)}$. Therefore the area of the surface of water is given by

$$A = \pi x^2 = \frac{\pi h}{a}.$$

(c) If the rate at which water is evaporating is proportional to the surface area, we have

$$\frac{dV}{dt} = -kA.$$

(The negative sign is included because the volume is decreasing.) By the chain rule, $\frac{dV}{dt} = \frac{dV}{dh} \cdot \frac{dh}{dt}$. We know $\frac{dV}{dh} = \frac{\pi h}{a}$ and $A = \frac{\pi h}{a}$ so

$$\frac{\pi h}{a} \frac{dh}{dt} = -k \frac{\pi h}{a} \quad \text{giving} \quad \frac{dh}{dt} = -k.$$

(d) Integrating gives

$$h = -kt + h_0.$$

Solving for t when $h = 0$ gives

$$t = \frac{h_0}{k}.$$

Solutions to Problems on Distribution Functions

1.

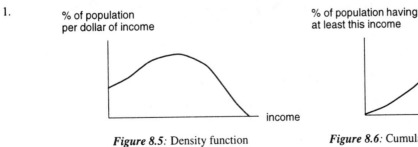

Figure 8.5: Density function

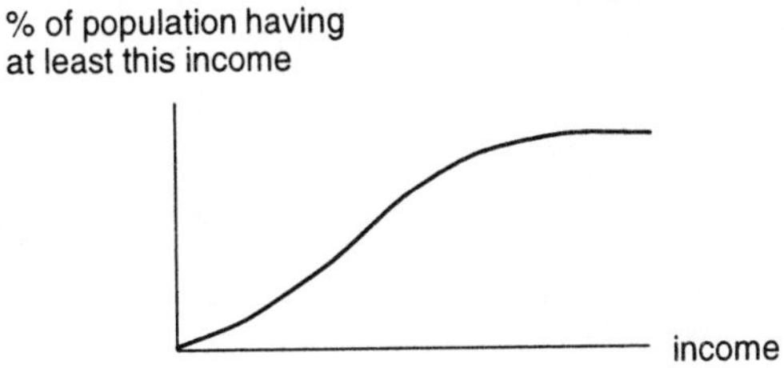

Figure 8.6: Cumulative distribution function

5. (a) Let $P(x)$ be the cumulative distribution function of the heights of the unfertilized plants. As do all cumulative distribution functions, $P(x)$ rises from 0 to 1 as x increases. The greatest number of plants will have heights in the range where $P(x)$ rises the most. The steepest rise appears to occur at about $x = 1$ m. Reading from the graph we see that $P(0.9) \approx 0.2$ and $P(1.1) \approx 0.8$, so that approximately $P(1.1) - P(0.9) = 0.8 - 0.2 = 0.6 = 60\%$ of the unfertilized plants grow to heights between 0.9 m and 1.1 m. Most of the plants grow to heights in the range 0.9 m to 1.1 m.

    (b) Let $P_A(x)$ be the cumulative distribution function of the plants that were fertilized with A. Since $P_A(x)$ rises the most in the range $0.7 \text{ m} \leq x \leq 0.9$ m, many of the plants fertilized with A will have heights in the range 0.7 m to 0.9 m. Reading from the graph of P_A, we find that $P_A(0.7) \approx 0.2$ and $P_A(0.9) \approx 0.8$, so $P_A(0.9) - P_A(0.7) \approx 0.8 - 0.2 = 0.6 = 60\%$ of the plants fertilized with A have heights between 0.7 m and 0.9 m. Fertilizer A had the effect of stunting the growth of the plants.

    On the other hand, the cumulative distribution function $P_B(x)$ of the heights of the plants fertilized with B rises the most in the range $1.1 \text{ m} \leq x \leq 1.3$ m, so most of these plants have heights in the range 1.1 m to 1.3 m. Fertilizer B caused the plants to grow about 0.2 m taller than they would have with no fertilizer.

9. (a) The two functions are shown below. The choice is based on the fact that the cumulative distribution does not decrease.

    (b) The cumulative distribution levels off to 1, so the top mark on the vertical scale must be 1.

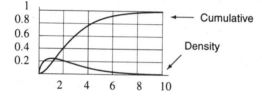

The total area under the density function must be 1. Since the area under the density function is about 2.5 boxes, each box must have area $1/2.5 = 0.4$. Since each box has a height of 0.2, the base must be 2.

13. (a) The fraction of maintenance checks completed in 15 minutes is $P(15) = 0.21$, or 21%.

    (b) Since $P(30) = 0.98$, we see that 98% of maintenance checks take 30 minutes or less. Therefore, only 2% take more than 30 minutes.

    (c) Since 8% take less than or equal to 10 minutes and 21% take less than or equal to 15 minutes, the fraction taking between 10 and 15 minutes must be $0.21 - 0.08 = 0.13$, or 13%.

    (d) We begin by making a table showing how the times are distributed. Reading from the table given in the problem, we see that the fraction of jobs completed between 0 and 5 minutes is 0.03, and the fraction completed between 5 and 10 minutes is 0.05. See Table 8.1.

TABLE 8.1   *Distribution of times for routine maintenance*

time period (minutes)	0-5	5-10	10-15	15-20	20-25	25-30	> 30
fraction of jobs	0.03	0.05	0.13	0.17	0.42	0.18	0.02

We now draw the histogram, arranging the vertical scale in such a way that the area of each bar in the histogram equals the fraction of jobs completed in the corresponding time period. For instance, since the first bar is to have area 0.03 and width 5 minutes, its height must be $0.03/5 = 0.006$. See Figure 8.7.

fraction of jobs
per minute

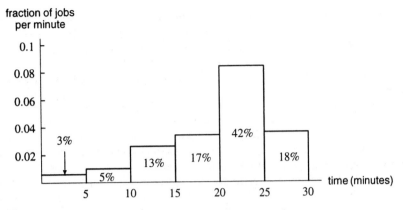

Figure 8.7: Histogram of maintenance times

(e) From Figure 8.7, we see that more of the jobs take between 20 and 25 minutes to complete, so this is the most likely length of time.

(f) The density function is a smoothed version of the histogram in Figure 8.7. Without more detailed information, we cannot know exactly how to draw it. A reasonable sketch is given in Figure 8.8.

(g) A graph is given in Figure 8.9. Since P is a cumulative distribution function, we know that $P(t)$ is approaching 1 as t gets large, but is never larger than 1.

fraction of maintenance checks
per minute spent

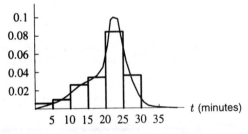

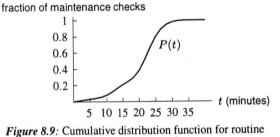

Figure 8.8: Density function for routine maintenance checks

Figure 8.9: Cumulative distribution function for routine maintenance checks

17. (a) The density function $f(r)$ will be zero outside the range $0 < r < 5$ and equal to a nonzero constant k inside this range. The area of the region under the density curve equals $5k$, which must equal 1, so $k = 0.2$. We have

$$f(r) = \begin{cases} 0 & \text{if } r \leq 0 \\ 0.2 & \text{if } 0 < r < 5 \\ 0 & \text{if } 5 \leq r. \end{cases}$$

The graph of $f(r)$ is given in Figure 8.10.

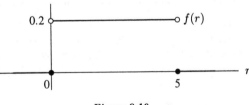

Figure 8.10

(b) The cumulative distribution function $F(r)$ equals the area of the region under the density function for $r < a$. From the graph in Figure 8.10 we see that the area is zero if $r < 0$; for $0 \leq r \leq 5$ the region is rectangular of height 0.2, width r, and area $0.2r$; and for $r > 5$ the area is 1. Thus

$$F(r) = \begin{cases} 0 & \text{if } r < 0 \\ 0.2r & \text{if } 0 \leq r \leq 5 \\ 1 & \text{if } 5 < r. \end{cases}$$

(c)   The cumulative distribution function $G(v)$ is the fraction of raindrops of volume less than or equal to v. Since volume $v = 4\pi r^3/3$, or equivalently $r = (3v/(4\pi))^{1/3}$, we see that $G(v)$ is the same as the fraction of raindrops of radius less than or equal to $(3v/(4\pi))^{1/3} = 0.62v^{1/3}$. In other words,

$$G(v) = F(0.62v^{1/3}) = \begin{cases} 0 & \text{if } 0.62v^{1/3} < 0 \\ (0.2)(0.62)v^{1/3} & \text{if } 0 \le 0.62v^{1/3} \le 5 \\ 1 & \text{if } 5 < 0.62v^{1/3}. \end{cases}$$

The final answer is thus

$$G(v) = \begin{cases} 0 & \text{if } v < 0 \\ 0.124v^{1/3} & \text{if } 0 \le v \le 523.6 \\ 1 & \text{if } 523.6 < v. \end{cases}$$

Note that the volume of a raindrop of radius 5 is $v = 4\pi 5^3/3 = 523.6$.

(d)   The density function $g(v)$ equals the derivative of the cumulative distribution function $G(v)$. We have

$$g(v) = \begin{cases} 0 & \text{if } v \le 0 \\ 0.0413v^{-2/3} & \text{if } 0 < v < 523.6 \\ 0 & \text{if } 523.6 \le v. \end{cases}$$

The density function is graphed in Figure 8.11. Notice that the density functions $f(r)$ for the radii of the raindrops and $g(v)$ for the volumes are quite different.

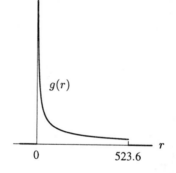

Figure 8.11

Solutions to Problems on Probability and Distributions

1.

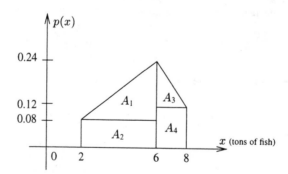

Splitting the figure into four pieces, we see that

$$\text{Area under the curve} = A_1 + A_2 + A_3 + A_4$$
$$= \frac{1}{2}(0.16)4 + 4(0.08) + \frac{1}{2}(0.12)2 + 2(0.12)$$
$$= 1.$$

We expect the area to be 1, since $\int_{-\infty}^{\infty} p(x)\,dx = 1$ for any probability density function, and $p(x)$ is 0 except when $2 \le x \le 8$.

5. (a) Since $\int_0^\infty p(x)\,dx = 1$, we have

$$1 = \int_0^\infty ae^{-0.122x}\,dx$$

$$= \frac{a}{-0.122}e^{-0.122x}\Big|_0^\infty = \frac{a}{0.122}.$$

So $a = 0.122$.

(b)

$$P(x) = \int_0^x p(t)\,dt$$

$$= \int_0^x 0.122e^{-0.122t}\,dt$$

$$= -e^{0.122t}\Big|_0^x = 1 - e^{-0.122x}.$$

(c) Median is the x such that

$$P(x) = 1 - e^{-0.122x} = 0.5.$$

So $e^{-0.122x} = 0.5$. Thus,

$$x = -\frac{\ln 0.5}{0.122} \approx 5.68 \text{ seconds}$$

and

$$\text{Mean} = \int_0^\infty x(0.122)e^{-0.122x}\,dx = -\int_0^\infty x\left(-0.122e^{-0.122x}\right)\,dx.$$

We now use integration by parts. Let $u = -x$ and $v' = -0.122e^{-0.122x}$. Then $u' = -1$, and $v = e^{-0.122x}$. Therefore,

$$\text{Mean} = -xe^{-0.122x}\Big|_0^\infty + \int_0^\infty e^{-0.122x}\,dx = \frac{1}{0.122} \approx 8.20 \text{ seconds}.$$

(d)

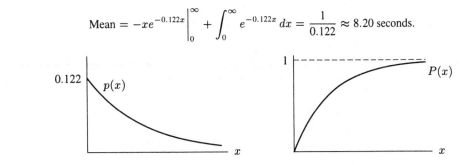

9. (a) First, we find the critical points of $p(x)$:

$$\frac{d}{dx}p(x) = \frac{1}{\sigma\sqrt{2\pi}}\left[\frac{-2(x-\mu)}{2\sigma^2}\right]e^{-\frac{(x-\mu)^2}{2\sigma^2}}$$

$$= -\frac{(x-\mu)}{\sigma^3\sqrt{2\pi}}e^{-\frac{(x-\mu)^2}{2\sigma^2}}.$$

This implies $x = \mu$ is the only critical point of $p(x)$.

To confirm that $p(x)$ is maximized at $x = \mu$, we rely on the first derivative test. As $-\frac{1}{\sigma^3\sqrt{2\pi}}e^{-\frac{(x-\mu)^2}{2\sigma^2}}$ is always negative, the sign of $p'(x)$ is the opposite of the sign of $(x-\mu)$; thus $p'(x) > 0$ when $x < \mu$, and $p'(x) < 0$ when $x > \mu$.

(b) To find the inflection points, we need to find where $p''(x)$ changes sign; that will happen only when $p''(x) = 0$. As

$$\frac{d^2}{dx^2}p(x) = -\frac{1}{\sigma^3\sqrt{2\pi}}e^{-\frac{(x-\mu)^2}{2\sigma^2}}\left[-\frac{(x-\mu)^2}{\sigma^2} + 1\right],$$

$p''(x)$ changes sign when $\left[-\frac{(x-\mu)^2}{\sigma^2} + 1\right]$ does, since the sign of the other factor is always negative. This occurs when

$$-\frac{(x-\mu)^2}{\sigma^2} + 1 = 0,$$
$$-(x-\mu)^2 = -\sigma^2,$$
$$x - \mu = \pm\sigma.$$

Thus, $x = \mu + \sigma$ or $x = \mu - \sigma$. Since $p''(x) > 0$ for $x < \mu - \sigma$ and $x > \mu + \sigma$ and $p''(x) < 0$ for $\mu - \sigma \le x \le \mu + \sigma$, these are in fact points of inflection.

(c) μ represents the mean of the distribution, while σ is the standard deviation. In other words, σ gives a measure of the "spread" of the distribution, i.e., how tightly the observations are clustered about the mean. A small σ tells us that most of the data are close to the mean; a large σ tells us that the data is spread out.

13. (a) P is the cumulative distribution function, so the percentage of the population that made between $20,000 and $50,000 is

$$P(50) - P(20) = 99\% - 75\% = 24\%.$$

Therefore $\frac{6}{25}$ of the population made between $20,000 and $50,000.

(b) The median income is the income such that half the people made less than this amount. Looking at the chart, we see that $P(12.6) = 50\%$, so the median must be $12,600.

(c) The cumulative distribution function looks something like this:

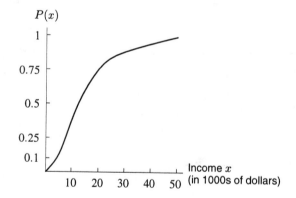

The density function is the derivative of the cumulative distribution. Qualitatively it looks like:

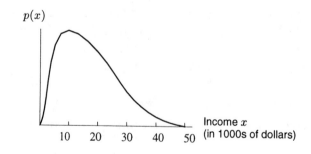

The density function has a maximum at about $8000. This means that more people have incomes around $8000 than around any other amount. On the density function, this is the highest point. On the cumulative distribution, this is the point of steepest slope (because $P' = p$), which is also the point of inflection.

CHAPTER NINE

1. Let $\dfrac{1}{1+x} = (1+x)^{-1}$. Then $f(0) = 1$.

$$
\begin{aligned}
f'(x) &= -1!(1+x)^{-2} & f'(0) &= -1, \\
f''(x) &= 2!(1+x)^{-3} & f''(0) &= 2!, \\
f'''(x) &= -3!(1+x)^{-4} & f'''(0) &= -3!, \\
f^{(4)}(x) &= 4!(1+x)^{-5} & f^{(4)}(0) &= 4!, \\
f^{(5)}(x) &= -5!(1+x)^{-6} & f^{(5)}(0) &= -5!, \\
f^{(6)}(x) &= 6!(1+x)^{-7} & f^{(6)}(0) &= 6!, \\
f^{(7)}(x) &= -7!(1+x)^{-8} & f^{(7)}(0) &= -7!, \\
f^{(8)}(x) &= 8!(1+x)^{-9} & f^{(8)}(0) &= 8!.
\end{aligned}
$$

$$
\begin{aligned}
P_4(x) &= 1 - x + x^2 - x^3 + x^4, \\
P_6(x) &= 1 - x + x^2 - x^3 + x^4 - x^5 + x^6, \\
P_8(x) &= 1 - x + x^2 - x^3 + x^4 - x^5 + x^6 - x^7 + x^8.
\end{aligned}
$$

5. Let $f(x) = \arctan x$. Then $f(0) = \arctan 0 = 0$, and

$$
\begin{aligned}
f'(x) &= 1/(1+x^2) = (1+x^2)^{-1} & f'(0) &= 1, \\
f''(x) &= (-1)(1+x^2)^{-2}2x & f''(0) &= 0, \\
f'''(x) &= 2!(1+x^2)^{-3}2^2x^2 + (-1)(1+x^2)^{-2}2 & f'''(0) &= -2, \\
f^{(4)}(x) &= -3!(1+x^2)^{-4}2^3x^3 + 2!(1+x^2)^{-3}2^3x \\
&\quad + 2!(1+x^2)^{-3}2^2x & f^{(4)}(0) &= 0.
\end{aligned}
$$

Therefore,

$$
P_3(x) = P_4(x) = x - \frac{1}{3}x^3.
$$

9. Let $f(x) = \dfrac{1}{\sqrt{1+x}} = (1+x)^{-1/2}$. Then $f(0) = 1$.

$$
\begin{aligned}
f'(x) &= -\tfrac{1}{2}(1+x)^{-3/2} & f'(0) &= -\tfrac{1}{2}, \\
f''(x) &= \tfrac{3}{2^2}(1+x)^{-5/2} & f''(0) &= \tfrac{3}{2^2}, \\
f'''(x) &= -\tfrac{3\cdot5}{2^3}(1+x)^{-7/2} & f'''(0) &= -\tfrac{3\cdot5}{2^3}, \\
f^{(4)}(x) &= \tfrac{3\cdot5\cdot7}{2^4}(1+x)^{-9/2} & f^{(4)}(0) &= \tfrac{3\cdot5\cdot7}{2^4}
\end{aligned}
$$

Then,

$$
\begin{aligned}
P_2(x) &= 1 - \frac{1}{2}x + \frac{1}{2!}\frac{3}{2^2}x^2 = 1 - \frac{1}{2}x + \frac{3}{8}x^2, \\
P_3(x) &= P_2(x) - \frac{1}{3!}\frac{3\cdot5}{2^3}x^3 = 1 - \frac{1}{2}x + \frac{3}{8}x^2 - \frac{5}{16}x^3, \\
P_4(x) &= P_3(x) + \frac{1}{4!}\frac{3\cdot5\cdot7}{2^4}x^4 = 1 - \frac{1}{2}x + \frac{3}{8}x^2 - \frac{5}{16}x^3 + \frac{35}{128}x^4.
\end{aligned}
$$

13.   Let $f(x) = \sin x$. $f(\frac{\pi}{2}) = 1$.

$$f'(x) = \cos x \qquad f'(\tfrac{\pi}{2}) = 0,$$
$$f''(x) = -\sin x \qquad f''(\tfrac{\pi}{2}) = -1,$$
$$f'''(x) = -\cos x \qquad f'''(\tfrac{\pi}{2}) = 0,$$
$$f^{(4)}(x) = \sin x \qquad f^{(4)}(\tfrac{\pi}{2}) = 1.$$

So,

$$P_4(x) = 1 + 0 - \frac{1}{2!}\left(x - \frac{\pi}{2}\right)^2 + 0 + \frac{1}{4!}\left(x - \frac{\pi}{2}\right)^4$$
$$= 1 - \frac{1}{2!}\left(x - \frac{\pi}{2}\right)^2 + \frac{1}{4!}\left(x - \frac{\pi}{2}\right)^4.$$

17.   Since $P_2(x)$ is the second degree Taylor polynomial for $f(x)$ about $x = 0$, $P_2(0) = f(0)$, which says $a = f(0)$. Since

$$\frac{d}{dx}P_2(x)\bigg|_{x=0} = f'(0),$$

$b = f'(0)$; and since

$$\frac{d^2}{dx^2}P_2(x)\bigg|_{x=0} = f''(0),$$

$2c = f''(0)$. In other words, a is the y-intercept of $f(x)$, b is the slope of the tangent line to $f(x)$ at $x = 0$ and c tells us the concavity of $f(x)$ near $x = 0$. So $c < 0$ since f is concave down; $b > 0$ since f is increasing; $a > 0$ since $f(0) > 0$.

21.

$$f(x) = \sin x \qquad f(\tfrac{\pi}{4}) = \tfrac{\sqrt{2}}{2},$$
$$f'(x) = \cos x \qquad f'(\tfrac{\pi}{4}) = \tfrac{\sqrt{2}}{2},$$
$$f''(x) = -\sin x \qquad f''(\tfrac{\pi}{4}) = -\tfrac{\sqrt{2}}{2},$$
$$f'''(x) = -\cos x \qquad f'''(\tfrac{\pi}{4}) = -\tfrac{\sqrt{2}}{2}.$$

$$\sin x = \frac{\sqrt{2}}{2} + \frac{\sqrt{2}}{2}\left(x - \frac{\pi}{4}\right) - \frac{\sqrt{2}}{2}\frac{(x - \frac{\pi}{4})^2}{2!} - \frac{\sqrt{2}}{2}\frac{(x - \frac{\pi}{4})^3}{3!} - \cdots$$
$$= \frac{\sqrt{2}}{2} + \frac{\sqrt{2}}{2}\left(x - \frac{\pi}{4}\right) - \frac{\sqrt{2}}{4}\left(x - \frac{\pi}{4}\right)^2 - \frac{\sqrt{2}}{12}\left(x - \frac{\pi}{4}\right)^3 - \cdots$$

25.   Let C_n be the coefficient of the nth term in the series. Note that

$$0 = C_1 = \frac{d}{dx}(x^2 e^{x^2})\bigg|_{x=0},$$

and since

$$\frac{1}{2} = C_6 = \frac{\frac{d^6}{dx^6}(x^2 e^{x^2})\big|_{x=0}}{6!},$$

we have

$$\frac{d^6}{dx^6}(x^2 e^{x^2})\bigg|_{x=0} = \frac{6!}{2} = 360.$$

29.   (a)   We'll make the following conjecture:
       "If $f(x)$ is a polynomial of degree n, i.e.

$$f(x) = a_0 + a_1 x + a_2 x^2 + \cdots + a_{n-1}x^{n-1} + a_n x^n,$$

       then $P_n(x)$, the nth degree Taylor polynomial for $f(x)$ about $x = 0$, is $f(x)$ itself."

(b) All we need to do is to calculate $P_n(x)$, the nth degree Taylor polynomial for f about $x = 0$ and see if it is the same as $f(x)$.

$$f(0) = a_0;$$
$$f'(0) = \left.(a_1 + 2a_2 x + \cdots + na_n x^{n-1})\right|_{x=0}$$
$$= a_1;$$
$$f''(0) = \left.(2a_2 + 3 \cdot 2a_3 x + \cdots + n(n-1)a_n x^{n-2})\right|_{x=0}$$
$$= 2!a_2.$$

If we continue doing this, we'll see in general

$$f^{(k)}(0) = k!a_k, \qquad k = 1, 2, 3, \cdots, n.$$

Therefore,

$$P_n(x) = f(0) + \frac{f'(0)}{1!}x + \frac{f''(0)}{2!}x^2 + \cdots + \frac{f^{(n)}(0)}{n!}x^n$$
$$= a_0 + a_1 x + a_2 x^2 + \cdots + a_n x^n$$
$$= f(x).$$

33. Let $f(x)$ be a function that has derivatives up to order n at $x = a$. Let

$$P_n(x) = C_0 + C_1(x - a) + \cdots + C_n(x - a)^n$$

be the polynomial of degree n that approximates $f(x)$ about $x = a$. We require that $P_n(x)$ and all of its first n derivatives agree with those of the function $f(x)$ at $x = a$, i.e., we want

$$f(a) = P_n(a),$$
$$f'(a) = P_n'(a),$$
$$f''(a) = P_n''(a),$$
$$\vdots$$
$$f^{(n)}(a) = P_n^{(n)}(a).$$

When we substitute $x = a$ in $P_n(x)$, all the terms except the first drop out, so

$$f(a) = C_0.$$

Now differentiate $P_n(x)$:

$$P_n'(x) = C_1 + 2C_2(x - a) + 3C_3(x - a)^2 + \cdots + nC_n(x - a)^{n-1}.$$

Substitute $x = a$ again, which yields

$$f'(a) = P_n'(a) = C_1.$$

Differentiate $P_n'(x)$:

$$P_n''(x) = 2C_2 + 3 \cdot 2C_3(x - a) + \cdots + n(n-1)C_n(x - a)^{n-2}$$

and substitute $x = a$ again:

$$f''(a) = P_n''(a) = 2C_2.$$

Differentiating and substituting again gives

$$f'''(a) = P_n'''(a) = 3 \cdot 2C_3.$$

Similarly,

$$f^{(k)}(a) = P_n{}^{(k)}(a) = k!C_k.$$

So, $C_0 = f(a)$, $C_1 = f'(a)$, $C_2 = \frac{f''(a)}{2!}$, $C_3 = \frac{f'''(a)}{3!}$, and so on.

If we adopt the convention that $f^{(0)}(a) = f(a)$ and $0! = 1$, then

$$C_k = \frac{f^{(k)}(a)}{k!}, \ k = 0, 1, 2, \cdots, n.$$

Therefore,

$$f(x) \approx P_n(x) = C_0 + C_1(x - a) + C_2(x - a)^2 \cdots + C_n(x - a)^n$$

$$= f(a) + f'(a)(x - a) + \frac{f''(a)}{2!}(x - a)^2 + \cdots + \frac{f^{(n)}(a)}{n!}(x - a)^n.$$

Solutions for Section 9.2

1. Yes.

5.

$$
\begin{aligned}
f(x) &= \tfrac{1}{1-x} = (1 - x)^{-1} & f(0) &= 1, \\
f'(x) &= -(1 - x)^{-2}(-1) = (1 - x)^{-2} & f'(0) &= 1, \\
f''(x) &= -2(1 - x)^{-3}(-1) = 2(1 - x)^{-3} & f''(0) &= 2, \\
f'''(x) &= -6(1 - x)^{-4}(-1) = 6(1 - x)^{-4} & f'''(0) &= 6.
\end{aligned}
$$

$$
\begin{aligned}
f(x) = \frac{1}{1 - x} &= 1 + 1 \cdot x + \frac{2x^2}{2!} + \frac{6x^3}{3!} + \cdots \\
&= 1 + x + x^2 + x^3 + \cdots
\end{aligned}
$$

9.

$$
\begin{aligned}
f(x) &= \tfrac{1}{x} & f(1) &= 1 \\
f'(x) &= -\tfrac{1}{x^2} & f'(1) &= -1 \\
f''(x) &= \tfrac{2}{x^3} & f''(1) &= 2 \\
f'''(x) &= -\tfrac{6}{x^4} & f'''(1) &= -6
\end{aligned}
$$

$$
\begin{aligned}
\frac{1}{x} &= 1 - (x - 1) + \frac{2(x - 1)^2}{2!} - \frac{6(x - 1)^3}{3!} + \cdots \\
&= 1 - (x - 1) + (x - 1)^2 - (x - 1)^3 + \cdots.
\end{aligned}
$$

13.

The graph suggests that the Taylor polynomials converge to $f(x) = \dfrac{1}{1 - x}$ on the interval $(-1, 1)$.

17. The coefficient of the nth term is $a_n = (-1)^{n+1}/n^2$. Now consider the ratio

$$\left| \frac{a_n}{a_{n+1}} \right| = \frac{(n+1)^2}{n^2} \to 1 = R \quad \text{as} \quad n \to \infty.$$

Thus, the radius of convergence is $R = 1$.

21. (a) Show that the sum of each group of fractions is more than $1/2$.
     (b) Explain why this shows that the harmonic series does not converge.

     (a) Notice that

$$\frac{1}{3} + \frac{1}{4} > \frac{1}{4} + \frac{1}{4} = \frac{2}{4} = \frac{1}{2}$$

$$\frac{1}{5} + \frac{1}{6} + \frac{1}{7} + \frac{1}{8} > \frac{1}{8} + \frac{1}{8} + \frac{1}{8} + \frac{1}{8} = \frac{4}{8} = \frac{1}{2}$$

$$\frac{1}{9} + \frac{1}{10} + \cdots + \frac{1}{16} > \frac{1}{16} + \frac{1}{16} + \cdots + \frac{1}{16} = \frac{8}{16} = \frac{1}{2}.$$

In the same way, we can see that the sum of the fractions in each grouping is greater than $1/2$.
     (b) Since the sum of the first n groups is greater than $n/2$, it follows that the harmonic series does not converge.

25. The partial sums are $S_0 = 1$, $S_1 = -1$, $S_2 = 2$, $S_9 = -5$, $S_{10} = 6$, $S_{99} = -50$, $S_{100} = 51$, $S_{999} = -500$, $S_{1000} = 501$, which appear to be oscillating further and further from 0. This series does not converge.

29. This is the series for $1/(1 - x)$ with x replaced by $1/4$, so the series converges to $1/(1 - (1/4)) = 4/3$.

Solutions for Section 9.3

1. We'll use

$$\sqrt{1 + y} = (1 + y)^{\frac{1}{2}} = 1 + \left(\frac{1}{2} \right) y + \left(\frac{1}{2} \right) \left(\frac{-1}{2} \right) \frac{y^2}{2!}$$

$$+ \left(\frac{1}{2} \right) \left(\frac{-1}{2} \right) \left(\frac{-3}{2} \right) \frac{y^3}{3!} + \cdots$$

$$= 1 + \frac{y}{2} - \frac{y^2}{8} + \frac{y^3}{16} - \cdots.$$

Substitute $y = -2x$.

$$\sqrt{1 - 2x} = 1 + \frac{(-2x)}{2} - \frac{(-2x)^2}{8} + \frac{(-2x)^3}{16} - \cdots$$

$$= 1 - x - \frac{x^2}{2} - \frac{x^3}{2} - \cdots$$

5. Substituting $x = -2y$ into $\ln(1 + x) = x - \frac{x^2}{2} + \frac{x^3}{3} - \frac{x^4}{4} + \cdots$ gives

$$\ln(1 - 2y) = (-2y) - \frac{(-2y)^2}{2} + \frac{(-2y)^3}{3} - \frac{(-2y)^4}{4} + \cdots$$

$$= -2y - 2y^2 - \frac{8}{3}y^3 - 4y^4 - \cdots.$$

9.

$$\frac{z}{e^{z^2}} = ze^{-z^2} = z \left(1 + (-z^2) + \frac{(-z^2)^2}{2!} + \frac{(-z^2)^3}{3!} + \cdots \right)$$

$$= z - z^3 + \frac{z^5}{2!} - \frac{z^7}{3!} + \cdots$$

13. The Taylor expansion about $\theta = 0$ for $\sin \theta$ is

$$\theta - \frac{\theta^3}{3!} + \frac{\theta^5}{5!} - \frac{\theta^7}{7!} + \cdots.$$

So

$$1 + \sin \theta = 1 + \theta - \frac{\theta^3}{3!} + \frac{\theta^5}{5!} - \frac{\theta^7}{7!} + \cdots.$$

The Taylor expansion about $\theta = 0$ for $\cos \theta$ is

$$\cos \theta = 1 - \frac{\theta^2}{2!} + \frac{\theta^4}{4!} - \frac{\theta^6}{6!} + \cdots.$$

The Taylor expansion for $\frac{1}{1+\theta}$ about $\theta = 0$ is

$$\frac{1}{1+\theta} = 1 - \theta + \theta^2 - \theta^3 + \theta^4 - \cdots.$$

So, substituting $-\theta^2$ for θ:

$$\frac{1}{1-\theta^2} = 1 - (-\theta^2) + (-\theta^2)^2 - (-\theta^2)^3 + (-\theta^2)^4 + \cdots$$
$$= 1 + \theta^2 + \theta^4 + \theta^6 + \theta^8 + \cdots.$$

For small θ, we can neglect the terms above quadratic in these expansions, giving:

$$1 + \sin \theta \approx 1 + \theta$$
$$\cos \theta \approx 1 - \frac{\theta^2}{2}$$
$$\frac{1}{1-\theta^2} \approx 1 + \theta^2.$$

For all $\theta \neq 0$, we have

$$1 - \frac{\theta^2}{2} < 1 + \theta^2.$$

Also, since $\theta^2 < \theta$ for $0 < \theta < 1$, we have

$$1 - \frac{\theta^2}{2} < 1 + \theta^2 < 1 + \theta.$$

So, for small positive θ, we have

$$\cos \theta < \frac{1}{1-\theta^2} < 1 + \sin \theta.$$

17.

$$\frac{1}{2+x} = \frac{1}{2(1+\frac{x}{2})} = \frac{1}{2}\left(1 + \frac{x}{2}\right)^{-1}$$
$$= \frac{1}{2}\left(1 - \frac{x}{2} + \left(\frac{x}{2}\right)^2 - \left(\frac{x}{2}\right)^3 + \cdots\right)$$

21. Using the binomial expansion we have

$$\sqrt{a^2 + x^2} = a\left(1 + \frac{x^2}{a^2}\right)^{1/2}$$
$$= a\left(1 + \frac{1}{2}\frac{x^2}{a^2} + \frac{(1/2)(-1/2)}{2!}\frac{x^4}{a^4} + \frac{(1/2)(-1/2)(-3/2)}{3!}\frac{x^6}{a^6} + \cdots\right)$$
$$= a\left(1 + \frac{1}{2}\frac{x^2}{a^2} - \frac{1}{8}\frac{x^4}{a^4} + \frac{1}{16}\frac{x^6}{a^6} + \cdots\right).$$

Similarly, we have

$$\sqrt{a^2 - x^2} = a\left(1 - \frac{1}{2}\frac{x^2}{a^2} - \frac{1}{8}\frac{x^4}{a^4} - \frac{1}{16}\frac{x^6}{a^6} - \cdots\right).$$

Combining gives

$$z = \sqrt{a^2 + x^2} - \sqrt{a^2 - x^2} = a\left(2\cdot\frac{1}{2}\frac{x^2}{a^2} + 2\cdot\frac{1}{16}\frac{x^6}{a^6} + \cdots\right) = \frac{x^2}{a} + \frac{1}{8}\frac{x^6}{a^5} + \cdots.$$

25. (a) $\mu = \frac{mM}{m+M}$.
     If $M \gg m$, then the denominator $m + M \approx M$, so $\mu \approx \frac{mM}{M} = m$.
     (b)

$$\mu = m\left(\frac{M}{m + M}\right) = m\left(\frac{\frac{1}{M}M}{\frac{m}{M} + \frac{M}{M}}\right) = m\left(\frac{1}{1 + \frac{m}{M}}\right)$$

We can use the binomial expansion since $\frac{m}{M} < 1$.

$$\mu = m\left[1 - \frac{m}{M} + \left(\frac{m}{M}\right)^2 - \left(\frac{m}{M}\right)^3 + \cdots\right]$$

(c) If $m \approx \frac{1}{1836}M$, then $\frac{m}{M} \approx \frac{1}{1836} \approx 0.000545$.
     So a first order approximation to μ would give $\mu = m(1 - 0.000545)$. The percentage difference from $\mu = m$ is -0.0545%.

29. (a) The Taylor series for $1/(1 - x) = 1 + x + x^2 + x^3 + \ldots$, so

$$\frac{1}{0.98} = \frac{1}{1 - 0.02} = 1 + (0.02) + (0.02)^2 + (0.02)^3 + \cdots$$
$$= 1.020408\ldots$$

(b) Since $d/dx(1/(1 - x)) = (1/(1 - x))^2$, the Taylor series for $1/(1 - x)^2$ is

$$\frac{d}{dx}(1 + x + x^2 + x^3 + \ldots) = 1 + 2x + 3x^2 + 4x^3 + \cdots$$

Thus

$$\frac{1}{(0.99)^2} = \frac{1}{(1 - 0.01)^2} = 1 + 2(0.01) + 3(0.0001) + 4(0.000001) + \cdots$$
$$= 1.0203040506\ldots$$

Solutions for Section 9.4

1. Yes, $a = 1$, ratio $= -1/2$.

5. No. Ratio between successive terms is not constant: $\frac{2x^2}{x} = 2x$, while $\frac{3x^3}{2x^2} = \frac{3}{2}x$.

9. No. Ratio between successive terms is not constant: $\frac{6z^2}{3z} = 2z$, while $\frac{9z^3}{6z^2} = \frac{3}{2}z$.

13. Sum $= \dfrac{1}{1 - (-y^2)} = \dfrac{1}{1 + y^2}, |y| < 1$.

17. $3 + \dfrac{3}{2} + \dfrac{3}{4} + \dfrac{3}{8}\cdots + \dfrac{3}{2^{10}} = 3\left(1 + \dfrac{1}{2} + \cdots + \dfrac{1}{2^{10}}\right) = \dfrac{3\left(1 - \frac{1}{2^{11}}\right)}{1 - \frac{1}{2}} = \dfrac{3\left(2^{11} - 1\right)}{2^{10}}$

21.   (a)

$$P_1 = 0$$
$$P_2 = 250(0.04)$$
$$P_3 = 250(0.04) + 250(0.04)^2$$
$$P_4 = 250(0.04) + 250(0.04)^2 + 250(0.04)^3$$
$$\vdots$$
$$P_n = 250(0.04) + 250(0.04)^2 + 250(0.04)^3 + \cdots + 250(0.04)^{n-1}$$

(b) $P_n = 250(0.04) \left(1 + (0.04) + (0.04)^2 + (0.04)^3 + \cdots + (0.04)^{n-2}\right) = 250\dfrac{0.04(1 - (0.04)^{n-1})}{1 - 0.04}$

(c)

$$P = \lim_{n \to \infty} P_n$$
$$= \lim_{n \to \infty} 250\frac{0.04(1 - (0.04)^{n-1})}{1 - 0.04}$$
$$= \frac{(250)(0.04)}{0.96} = 0.04Q \approx 10.42$$

Thus, $\lim\limits_{n \to \infty} P_n = 10.42$ and $\lim\limits_{n \to \infty} Q_n = 260.42$. We would expect these limits to differ because one is right before taking a tablet, one is right after. We would expect the difference between them to be 250 mg, the amount of ampicillin in one tablet.

25.   (a)

$$\text{Total amount of money deposited} = 100 + 92 + 84.64 + \cdots$$
$$= 100 + 100(0.92) + 100(0.92)^2 + \cdots$$
$$= \frac{100}{1 - 0.92} = 1250 \quad \text{dollars}$$

(b)   Credit multiplier $= 1250/100 = 12.50$
The 12.50 is the factor by which the bank has increased its deposits, from $100 to $1250.

Solutions for Section 9.5

1.   No, a Fourier series has terms of the form $\cos nx$, not $\cos^n x$.

5.

$$a_0 = \frac{1}{2\pi} \int_{-\pi}^{\pi} f(x)\, dx = \frac{1}{2\pi} \left[\int_{-\pi}^{0} -1\, dx + \int_{0}^{\pi} 1\, dx \right] = 0$$

$$a_1 = \frac{1}{\pi} \int_{-\pi}^{\pi} f(x) \cos x\, dx = \frac{1}{\pi} \left[\int_{-\pi}^{0} -\cos x\, dx + \int_{0}^{\pi} \cos x\, dx \right]$$
$$= \frac{1}{\pi} \left[-\sin x \Big|_{-\pi}^{0} + \sin x \Big|_{0}^{\pi} \right] = 0.$$

Similarly, a_2 and a_3 are both 0.
(In fact, notice $f(x) \cos nx$ is an odd function, so $\int_{-\pi}^{\pi} f(x) \cos nx = 0$.)

$$b_1 = \frac{1}{\pi} \int_{-\pi}^{\pi} f(x) \sin x\, dx = \frac{1}{\pi} \left[\int_{-\pi}^{0} -\sin x\, dx + \int_{0}^{\pi} \sin x\, dx \right]$$
$$= \frac{1}{\pi} \left[\cos x \Big|_{-\pi}^{0} + (-\cos x) \Big|_{0}^{\pi} \right] = \frac{4}{\pi}.$$

$$b_2 = \frac{1}{\pi} \int_{-\pi}^{\pi} f(x) \sin 2x \, dx = \frac{1}{\pi} \left[\int_{-\pi}^{0} -\sin 2x \, dx + \int_{0}^{\pi} \sin 2x \, dx \right]$$

$$= \frac{1}{\pi} \left[\frac{1}{2} \cos 2x \Big|_{-\pi}^{0} + \left(-\frac{1}{2} \cos 2x\right) \Big|_{0}^{\pi} \right] = 0.$$

$$b_3 = \frac{1}{\pi} \int_{-\pi}^{\pi} f(x) \sin 3x \, dx = \frac{1}{\pi} \left[\int_{-\pi}^{0} -\sin 3x \, dx + \int_{0}^{\pi} \sin 3x \, dx \right]$$

$$= \frac{1}{\pi} \left[\frac{1}{3} \cos 3x \Big|_{-\pi}^{0} + \left(-\frac{1}{3} \cos 3x\right) \Big|_{0}^{\pi} \right] = \frac{4}{3\pi}.$$

Thus, $F_1(x) = F_2(x) = \frac{4}{\pi} \sin x$ and $F_3(x) = \frac{4}{\pi} \sin x + \frac{4}{3\pi} \sin 3x$.

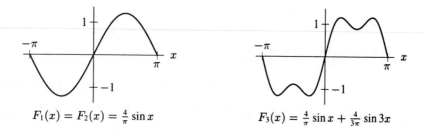

$$F_1(x) = F_2(x) = \frac{4}{\pi} \sin x \qquad\qquad F_3(x) = \frac{4}{\pi} \sin x + \frac{4}{3\pi} \sin 3x$$

9.

$$a_0 = \frac{1}{2\pi} \int_{-\pi}^{\pi} h(x) \, dx = \frac{1}{2\pi} \int_{0}^{\pi} x \, dx = \frac{\pi}{4}$$

As in Problem 10, we use the integral table (III-15 and III-16) to find formulas for a_n and b_n.

$$a_n = \frac{1}{\pi} \int_{-\pi}^{\pi} h(x) \cos(nx) \, dx = \frac{1}{\pi} \int_{0}^{\pi} x \cos nx \, dx = \frac{1}{\pi} \left(\frac{x}{n} \sin(nx) + \frac{1}{n^2} \cos(nx) \right) \Big|_{0}^{\pi}$$

$$= \frac{1}{\pi} \left(\frac{1}{n^2} \cos(n\pi) - \frac{1}{n^2} \right)$$

$$= \frac{1}{n^2 \pi} \left(\cos(n\pi) - 1 \right).$$

Note that since $\cos(n\pi) = (-1)^n$, $a_n = 0$ if n is even and $a_n = -\frac{2}{n^2\pi}$ if n is odd.

$$b_n = \frac{1}{\pi} \int_{-\pi}^{\pi} h(x) \cos(nx) \, dx = \frac{1}{\pi} \int_{0}^{\pi} x \sin x \, dx$$

$$= \frac{1}{\pi} \left(-\frac{x}{n} \cos(nx) + \frac{1}{n^2} \sin(nx) \right) \Big|_{0}^{\pi}$$

$$= \frac{1}{\pi} \left(-\frac{\pi}{n} \cos(n\pi) \right)$$

$$= -\frac{1}{n} \cos(n\pi)$$

$$= \frac{1}{n} (-1)^{n+1} \quad \text{if } n \geq 1$$

We have that the nth Fourier polynomial for h (for $n \geq 1$) is

$$H_n(x) = \frac{\pi}{4} + \sum_{i=1}^{n} \left(\frac{1}{i^2 \pi} \left(\cos(i\pi) - 1 \right) \cdot \cos(ix) + \frac{(-1)^{i+1} \sin(ix)}{i} \right).$$

This can also be written as

$$H_n(x) = \frac{\pi}{4} + \sum_{i=1}^{n} \frac{(-1)^{i+1} \sin(ix)}{i} + \sum_{i=1}^{\left[\frac{n}{2}\right]} \frac{-2}{(2i-1)^2\pi} \cos((2i-1)x)$$

where $\left[\frac{n}{2}\right]$ denotes the biggest integer smaller than or equal to $\frac{n}{2}$. In particular, we have the following graphs:

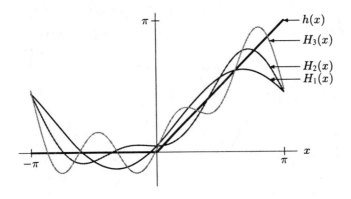

13. Since the period is 2, we make the substitution $t = \pi x - \pi$. Thus, $x = \frac{t+\pi}{\pi}$. We find the Fourier coefficients. Notice that all of the integrals are the same as in Problem 12 except for an extra factor of 2. Thus, $a_0 = 1$, $a_n = 0$, and $b_n = \frac{4}{\pi n}(-1)^{n+1}$, so:

$$G_4(t) = 1 + \frac{4}{\pi} \sin t - \frac{2}{\pi} \sin 2t + \frac{4}{3\pi} \sin 3t - \frac{1}{\pi} \sin 4t.$$

Again, we substitute back in to get a Fourier polynomial in terms of x:

$$F_4(x) = 1 + \frac{4}{\pi} \sin(\pi x - \pi) - \frac{2}{\pi} \sin(2\pi x - 2\pi)$$

$$+ \frac{4}{3\pi} \sin(3\pi x - 3\pi) - \frac{1}{\pi} \sin(4\pi x - 4\pi)$$

$$= 1 - \frac{4}{\pi} \sin(\pi x) - \frac{2}{\pi} \sin(2\pi x) - \frac{4}{3\pi} \sin(3\pi x) - \frac{1}{\pi} \sin(4\pi x).$$

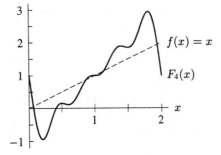

Notice in this case, the terms in our series are $\sin(n\pi x)$, not $\sin(2\pi nx)$, as in Problem 12. In general, the terms will be $\sin(n\frac{2\pi}{b}x)$, where b is the period.

17. Since each square in the graph has area $\left(\frac{\pi}{4}\right) \cdot (0.2)$,

$$a_0 = \frac{1}{2\pi} \int_{-\pi}^{\pi} f(x)\,dx$$

$$= \frac{1}{2\pi} \cdot \left(\frac{\pi}{4}\right) \cdot (0.2) \left[\text{Number of squares under graph above } x\text{-axis}\right.$$

$$\left. - \text{Number of squares above graph below } x \text{ axis}\right]$$

$$\approx \frac{1}{2\pi} \cdot \left(\frac{\pi}{4}\right) \cdot (0.2) \cdot [13 + 11 - 14] = 0.25.$$

Approximate the Fourier coefficients using Riemann sums.

$$a_1 = \frac{1}{\pi} \int_{-\pi}^{\pi} f(x) \cos x \, dx$$

$$\approx \frac{1}{\pi} \left[f(-\pi) \cos(-\pi) + f\left(-\frac{\pi}{2}\right) \cos\left(-\frac{\pi}{2}\right) + f(0) \cos(0) + f\left(\frac{\pi}{2}\right) \cos\left(\frac{\pi}{2}\right) \right] \cdot \frac{\pi}{2}$$

$$= \frac{1}{\pi} \left[(0.92)(-1) + (1)(0) + (-1.7)(1) + (0.7)(0) \right] \cdot \frac{\pi}{2}$$

$$= -1.31$$

Similarly for b_1:

$$b_1 = \frac{1}{\pi} \int_{-\pi}^{\pi} f(x) \sin x \, dx$$

$$\approx \frac{1}{\pi} \left[f(-\pi) \sin(-\pi) + f\left(-\frac{\pi}{2}\right) \sin\left(-\frac{\pi}{2}\right) + f(0) \sin(0) + f\left(\frac{\pi}{2}\right) \sin\left(\frac{\pi}{2}\right) \right] \cdot \frac{\pi}{2}$$

$$= \frac{1}{\pi} \left[(0.92)(0) + (1)(-1) + (-1.7)(0) + (0.7)(1) \right] \cdot \frac{\pi}{2}$$

$$= -0.15.$$

So our first Fourier approximation is

$$F_1(x) = 0.25 - 1.31 \cos x - 0.15 \sin x.$$

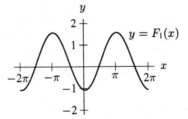

Similarly for a_2:

$$a_2 = \frac{1}{\pi} \int_{-\pi}^{\pi} f(x) \cos 2x \, dx$$

$$\approx \frac{1}{\pi} \left[f(-\pi) \cos(-2\pi) + f\left(-\frac{\pi}{2}\right) \cos(-\pi) + f(0) \cos(0) + f\left(\frac{\pi}{2}\right) \cos(-\pi) \right] \cdot \frac{\pi}{2}$$

$$= \frac{1}{\pi} \left[(0.92)(1) + (1)(-1) + (-1.7)(1) + (0.7)(-1) \right] \cdot \frac{\pi}{2}$$

$$= -1.24$$

Similarly for b_2:

$$b_2 = \frac{1}{\pi} \int_{-\pi}^{\pi} f(x) \sin 2x \, dx$$

$$\approx \frac{1}{\pi} \left[f(-\pi) \sin(-2\pi) + f\left(-\frac{\pi}{2}\right) \sin(-\pi) + f(0) \sin(0) + f\left(\frac{\pi}{2}\right) \sin(-\pi) \right] \cdot \frac{\pi}{2}$$

$$= \frac{1}{\pi} \left[(0.92)(0) + (1)(0) + (-1.7)(0) + (0.7)(0) \right] \cdot \frac{\pi}{2}$$

$$= 0.$$

So our second Fourier approximation is

$$F_2(x) = 0.25 - 1.31 \cos x - 0.15 \sin x - 1.24 \cos 2x.$$

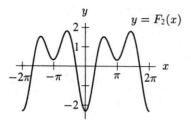

As you can see from comparing our graphs of F_1 and F_2 to the original, our estimates of the Fourier coefficients are not very accurate.

There are other methods of estimating the Fourier coefficients such as taking other Riemann sums, using Simpson's rule, and using the trapezoid rule. With each method, the greater the number of subdivisions, the more accurate the estimates of the Fourier coefficients.

The actual function graphed in the problem was

$$y = \frac{1}{4} - 1.3\cos x - \frac{\sin(\frac{3}{5})}{\pi}\sin x - \frac{2}{\pi}\cos 2x - \frac{\cos 1}{3\pi}\sin 2x$$

$$= 0.25 - 1.3\cos x - 0.18\sin x - 0.63\cos 2x - 0.057\sin 2x.$$

21.   (a)

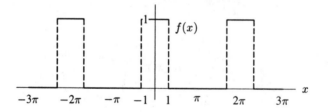

The energy of the pulse train f is

$$E = \frac{1}{\pi}\int_{-\pi}^{\pi}(f(x))^2\,dx = \frac{1}{\pi}\int_{-1}^{1}1^2 = \frac{1}{\pi}(1-(-1)) = \frac{2}{\pi}.$$

Next, find the Fourier coefficients:

$$a_0 = \text{average value of } f \text{ on } [-\pi, \pi] = \frac{1}{2\pi}(\text{ Area}) = \frac{1}{2\pi}(2) = \frac{1}{\pi},$$

$$a_k = \frac{1}{\pi}\int_{-\pi}^{\pi}f(x)\cos kx\,dx = \frac{1}{\pi}\int_{-1}^{1}\cos kx\,dx = \frac{1}{k\pi}\sin kx\Big|_{-1}^{1}$$

$$= \frac{1}{k\pi}(\sin k - \sin(-k)) = \frac{1}{k\pi}(2\sin k),$$

$$b_k = \frac{1}{\pi}\int_{-\pi}^{\pi}f(x)\sin kx\,dx = \frac{1}{\pi}\int_{-1}^{1}\sin kx\,dx = -\frac{1}{k\pi}\cos kx\Big|_{-1}^{1}$$

$$= -\frac{1}{k\pi}(\cos k - \cos(-k)) = \frac{1}{k\pi}(0) = 0.$$

The energy of f contained in the constant term is

$$A_0^2 = 2a_0^2 = 2\left(\frac{1}{\pi}\right)^2 = \frac{2}{\pi^2}$$

which is

$$\frac{A_0^2}{E} = \frac{2/\pi^2}{2/\pi} = \frac{1}{\pi} \approx 0.3183 = 31.83\% \quad \text{of the total.}$$

The fraction of energy contained in the first harmonic is

$$\frac{A_1^2}{E} = \frac{a_1^2}{E} = \frac{\left(\frac{2\sin 1}{\pi}\right)^2}{\frac{2}{\pi}} \approx 0.4508 = 45.08\%.$$

The fraction of energy contained in both the constant term and the first harmonic together is

$$\frac{A_0^2}{E} + \frac{A_1^2}{E} \approx 0.7691 = 76.91\%.$$

(b) The fraction of energy contained in the second harmonic is

$$\frac{A_2^2}{E} = \frac{a_2^2}{E} = \frac{\left(\frac{\sin 2}{\pi}\right)^2}{\frac{2}{\pi}} \approx 0.1316 = 13.16\%$$

so the fraction of energy contained in the constant term and first two harmonics is

$$\frac{A_0^2}{E} + \frac{A_1^2}{E} + \frac{A_2^2}{E} \approx 0.7691 + 0.1316 = 0.9007 = 90.07\%.$$

Therefore, the constant term and the first two harmonics are needed to capture 90% of the energy of f.

(c)

$$F_3(x) = \frac{1}{\pi} + \frac{2\sin 1}{\pi}\cos x + \frac{\sin 2}{\pi}\cos 2x + \frac{2\sin 3}{3\pi}\cos 3x$$

25. The easiest way to do this is to use Problem 24.

$$\int_{-\pi}^{\pi} \sin^2 mx\, dx = \int_{-\pi}^{\pi}(1-\cos^2 mx)\,dx = \int_{-\pi}^{\pi} dx - \int_{-\pi}^{\pi}\cos^2 mx\, dx$$
$$= 2\pi - \pi \quad \text{using Problem 24}$$
$$= \pi.$$

Solutions for Chapter 9 Review

1. $e^x \approx 1 + e(x-1) + \frac{e}{2}(x-1)^2$

5.

$$\theta^2\cos\theta^2 = \theta^2\left(1 - \frac{(\theta^2)^2}{2!} + \frac{(\theta^2)^4}{4!} - \frac{(\theta^2)^6}{6!} + \cdots\right)$$
$$= \theta^2 - \frac{\theta^6}{2!} + \frac{\theta^{10}}{4!} - \frac{\theta^{14}}{6!} + \cdots$$

9.

$$\frac{a}{a+b} = \frac{a}{a(1+\frac{b}{a})} = \left(1+\frac{b}{a}\right)^{-1} = 1 - \frac{b}{a} + \left(\frac{b}{a}\right)^2 - \left(\frac{b}{a}\right)^3 + \cdots$$

13. Factoring out a 3, we see

$$3\left(1 + 1 + \frac{1}{2!} + \frac{1}{3!} + \frac{1}{4!} + \frac{1}{5!} + \cdots\right) = 3e^1 = 3e.$$

17. First we use the Taylor series expansion for $\ln(1+t)$,

$$\ln(1+t) = t - \frac{1}{2}t^2 + \frac{1}{3}t^3 - \frac{1}{4}t^4 + \cdots$$

to find the Taylor series expansion of $\ln(1+x+x^2)$ by putting $t = x + x^2$. We get

$$\ln(1+x+x^2) = x + \frac{1}{2}x^2 - \frac{2}{3}x^3 + \frac{1}{4}x^4 + \cdots.$$

Next we use the Taylor series for $\sin x$ to get

$$\sin^2 x = (\sin x)^2 = (x - \frac{1}{6}x^3 + \frac{1}{120}x^5 - \cdots)^2 = x^2 - \frac{1}{3}x^4 + \cdots.$$

Finally,

$$\frac{\ln(1+x+x^2) - x}{\sin^2 x} = \frac{\frac{1}{2}x^2 - \frac{2}{3}x^3 + \frac{1}{4}x^4 + \cdots}{x^2 - \frac{1}{3}x^4 + \cdots} \rightarrow \frac{1}{2}, \quad \text{as} \quad x \rightarrow 0.$$

21. (a) To find when V takes on its minimum values, set $\frac{dV}{dr} = 0$. So

$$-V_0\frac{d}{dr}\left(2\left(\frac{r_0}{r}\right)^6 - \left(\frac{r_0}{r}\right)^{12}\right) = 0$$

$$-V_0\left(-12r_0^6 r^{-7} + 12r_0^{12}r^{-13}\right) = 0$$

$$12r_0^6 r^{-7} = 12r_0^{12}r^{-13}$$

$$r_0^6 = r^6$$

$$r = r_0.$$

Rewriting $V'(r)$ as $\frac{12r_0^6 V_0}{r^7}\left(1 - \left(\frac{r_0}{r}\right)^6\right)$, we see that $V'(r) > 0$ for $r > r_0$ and $V'(r) < 0$ for $r < r_0$. Thus, $V = -V_0(2(1)^6 - (1)^{12}) = -V_0$ is a minimum.

(Note: We discard the negative root $-r_0$ since the distance r must be positive.)

(b)

$$V(r) = -V_0\left(2\left(\frac{r_0}{r}\right)^6 - \left(\frac{r_0}{r}\right)^{12}\right) \qquad V(r_0) = -V_0$$

$$V'(r) = -V_0(-12r_0^6 r^{-7} + 12r_0^{12}r^{-13}) \qquad V'(r_0) = 0$$

$$V''(r) = -V_0(84r_0^6 r^{-8} - 156r_0^{12}r^{-14}) \qquad V''(r_0) = 72V_0 r_0^{-2}$$

The Taylor series is thus:

$$V(r) = -V_0 + 72V_0 r_0^{-2} \cdot (r - r_0)^2 \cdot \frac{1}{2} + \cdots$$

(c) The difference between V and its minimum value $-V_0$ is

$$V - (-V_0) = 36V_0\frac{(r - r_0)^2}{r_0^2} + \cdots$$

which is approximately proportional to $(r - r_0)^2$ since terms containing higher powers of $(r - r_0)$ have relatively small values for r near r_0.

(d) From part (a) we know that $dV/dr = 0$ when $r = r_0$, hence $F = 0$ when $r = r_0$. Since, if we discard powers of $(r - r_0)$ higher than the second,

$$V(r) \approx -V_0 \left(1 - 36\frac{(r - r_0)^2}{r_0^2}\right)$$

giving

$$F = -\frac{dV}{dr} \approx 72 \cdot \frac{r - r_0}{r_0^2}(-V_0) = -72V_0\frac{r - r_0}{r_o^2}.$$

So F is approximately proportional to $(r - r_0)$.

25. Since expanding $f(x + h)$ and $g(x + h)$ in Taylor series gives

$$f(x + h) = f(x) + f'(x)h + \frac{f''(x)}{2!}h^2 + \cdots,$$

$$g(x + h) = g(x) + g'(x)h + \frac{f''(x)}{2!}h^2 + \cdots,$$

we substitute to get

$$\frac{f(x + h)g(x + h) - f(x)g(x)}{h}$$

$$= \frac{(f(x) + f'(x)h + \frac{1}{2}f''(x)h^2 + \cdots)(g(x) + g'(x)h + \frac{1}{2}g''(x)h^2 + \cdots) - f(x)g(x)}{h}$$

$$= \frac{f(x)g(x) + (f'(x)g(x) + f(x)g'(x))h + \text{ Terms in } h^2 \text{ and higher powers} - f(x)g(x)}{h}$$

$$= \frac{h(f'(x)g(x) + f(x)g'(x) + \text{ Terms in } h \text{ and higher powers})}{h}$$

$$= f'(x)g(x) + f(x)g'(x) + \text{ Terms in } h \text{ and higher powers.}$$

Thus, taking the limit as $h \to 0$, we get

$$\frac{d}{dx}(f(x)g(x)) = \lim_{h \to 0} \frac{f(x + h)g(x + h) - f(x)g(x)}{h}$$

$$= f'(x)g(x) + f(x)g'(x).$$

29. Let us begin by finding the Fourier coefficients for $f(x)$. Since f is odd, $\int_{-\pi}^{\pi} f(x)\,dx = 0$ and $\int_{-\pi}^{\pi} f(x)\cos nx\,dx = 0$. Thus $a_i = 0$ for all $i \geq 0$. On the other hand,

$$b_i = \frac{1}{\pi} \int_{-\pi}^{\pi} f(x)\sin nx\,dx = \frac{1}{\pi}\left[\int_{-\pi}^{0} -\sin(nx)\,dx + \int_{0}^{\pi} \sin(nx)\,dx\right]$$

$$= \frac{1}{\pi}\left[\frac{1}{n}\cos(nx)\Big|_{-\pi}^{0} - \frac{1}{n}\cos(nx)\Big|_{0}^{\pi}\right]$$

$$= \frac{1}{n\pi}\left[\cos 0 - \cos(-n\pi) - \cos(n\pi) + \cos 0\right]$$

$$= \frac{2}{n\pi}\left(1 - \cos(n\pi)\right).$$

Since $\cos(n\pi) = (-1)^n$, this is 0 if n is even, and $\frac{4}{n\pi}$ if n is odd. Thus the nth Fourier polynomial (where n is odd) is

$$F_n(x) = \frac{4}{\pi}\sin x + \frac{4}{3\pi}\sin 3x + \cdots + \frac{4}{n\pi}\sin(nx).$$

As $n \to \infty$, the nth Fourier polynomial must approach $f(x)$ on the interval $(-\pi, \pi)$, except at the point $x = 0$ (where f is not continuous). In particular, if $x = \frac{\pi}{2}$,

$$F_n(1) = \frac{4}{\pi}\sin\frac{\pi}{2} + \frac{4}{3\pi}\sin\frac{3\pi}{2} + \frac{4}{5\pi}\sin\frac{5\pi}{2} + \frac{4}{7\pi}\sin\frac{7\pi}{2} + \cdots + \frac{4}{n\pi}\sin\frac{n\pi}{2}$$

$$= \frac{4}{\pi}\left(1 - \frac{1}{3} + \frac{1}{5} - \frac{1}{7} + \cdots + (-1)^{2n+1}\frac{1}{2n + 1}\right).$$

But $F_n(1)$ approaches $f(\frac{\pi}{2}) = 1$ as $n \to \infty$, so

$$\frac{\pi}{4} F_n(1) = 1 - \frac{1}{3} + \frac{1}{5} - \frac{1}{7} + \cdots + (-1)^{2n+1} \frac{1}{2n+1} \to \frac{\pi}{4} \cdot 1 = \frac{\pi}{4}.$$

33. Since $g(x) = f(x + c)$, we have that $[g(x)]^2 = [f(x + c)]^2$, so g^2 is f^2 shifted horizontally by c. Since f has period 2π, so does f^2 and g^2. If you think of the definite integral as an area, then because of the periodicity, integrals of f^2 over any interval of length 2π have the same value. So

$$\text{Energy of } f = \int_{-\pi}^{\pi} (f(x))^2 \, dx = \int_{-\pi+c}^{\pi+c} (f(x))^2 \, dx.$$

Now we know that

$$\text{Energy of } g = \frac{1}{\pi} \int_{-\pi}^{\pi} (g(x))^2 \, dx$$

$$= \frac{1}{\pi} \int_{-\pi}^{\pi} (f(x + c))^2 \, dx.$$

Using the substitution $t = x + c$, we see that the two energies are equal.

Solutions to Problems on Convergence Theorems

1. Let S_n be the nth partial sum for $\sum a_n$ and let T_n be the nth partial sum for $\sum b_n$. Then the nth partial sums for $\sum(a_n + b_n)$, $\sum(a_n - b_n)$, and $\sum ka_n$ are $S_n + T_n$, $S_n - T_n$, and kS_n, respectively. To show that these series converge, we have to show that the limits of their partial sums exist. By the properties of limits,

$$\lim_{n \to \infty} (S_n + T_n) = \lim_{n \to \infty} S_n + \lim_{n \to \infty} T_n$$
$$\lim_{n \to \infty} (S_n - T_n) = \lim_{n \to \infty} S_n - \lim_{n \to \infty} T_n$$
$$\lim_{n \to \infty} kS_n = k \lim_{n \to \infty} S_n.$$

This proves that the limits of the partial sums exist, so the series converge.

5. Since $\ln n \leq n$ for $n \geq 2$, we have $1/\ln n \geq 1/n$, so the series diverges by comparison with the harmonic series, $\sum 1/n$.

9. Using right hand approximating sums or lower sums for the integral of $f(x) = x^{-p}$ over the interval $1 \leq x \leq n$ with uniform subdivisions of width 1 gives:

$$\frac{1}{2^p} + \cdots + \frac{1}{n^p} \leq \int_1^n x^{-p} \, dx = \left. \frac{1}{1-p} x^{1-p} \right|_1^n = \frac{1}{1-p}(n^{1-p} - 1).$$

Since $p > 1$, the exponent $1 - p$ is negative, so as $n \to \infty$, the sequence of partial sums converges. Hence the series converges.

Solutions to Problems on the Error in Taylor Approximations

1. (a) The Taylor polynomial of degree 0 about $t = 0$ for $f(t) = e^t$ is simply $P_0(x) = 1$. Since $e^t \geq 1$ on $[0, 0.5]$, the approximation is an underestimate.

(b) Using the zero degree error bound, if $|f'(t)| \leq M$ for $0 \leq t \leq 0.5$, then

$$|E_0| \leq M \cdot |t| \leq M(0.5).$$

Since $|f'(t)| = |e^t| = e^t$ is increasing on $[0, 0.5]$,

$$|f'(t)| \leq e^{0.5} < \sqrt{4} = 2.$$

Therefore

$$|E_0| \leq (2)(0.5) = 1.$$

(Note: By looking at a graph of $f(t)$ and its 0th degree approximation, it is easy to see that the greatest error occurs when $t = 0.5$, and the error is $e^{0.5} - 1 \approx 0.65 < 1$. So our error bound works.)

5. Let $f(x) = \tan x$. The error bound for the Taylor approximation of degree three for $f(1) = \tan 1$ about $x = 0$ is:

$$|E_3| = |f(1) - P_3(x)| \leq \frac{M \cdot |1 - 0|^4}{4!} = \frac{M}{24}$$

where $|f^{(4)}(x)| \leq M$ for $0 \leq x \leq 1$. Now, $f^{(4)}(x) = \frac{16 \sin x}{\cos^3 x} + \frac{24 \sin^3 x}{\cos^5 x}$. From a graph of $f^{(4)}(x)$, we see that $f^{(4)}(x)$ is increasing for x between 0 and 1. Thus,

$$|f^{(4)}(x)| \leq |f^{(4)}(1)| \approx 396,$$

so

$$|E_3| \leq \frac{396}{24} = 16.5.$$

This is not a very helpful error bound! The reason the error bound is so huge is that $x = 1$ is getting near the vertical asymptote of the tangent graph, and the fourth derivative is enormous there.

9. The maximum possible error for the nth degree Taylor polynomial about $x = 0$ approximating $\cos x$ is $|E_n| \leq \frac{M \cdot |x - 0|^{n+1}}{(n+1)!}$, where $|\cos^{(n+1)} x| \leq M$ for $0 \leq x \leq 1$. Now the derivatives of $\cos x$ are simply $\cos x, \sin x, -\cos x$, and $-\sin x$. The largest magnitude these ever take is 1, so $|\cos^{(n+1)}(x)| \leq 1$, and thus $|E_n| \leq \frac{|x|^{n+1}}{(n+1)!} \leq \frac{1}{(n+1)!}$. The same argument works for $\sin x$.

13. (a)

TABLE 9.1 $E_1 = \sin x - x$		
x	$\sin x$	E
-0.5	-0.4794	0.0206
-0.4	-0.3894	0.0106
-0.3	-0.2955	0.0045
-0.2	-0.1987	0.0013
-0.1	-0.0998	0.0002

TABLE 9.2 $E_1 = \sin x - x$		
x	$\sin x$	E
0	0	0
0.1	0.0998	-0.0002
0.2	0.1987	-0.0013
0.3	0.2955	-0.0045
0.4	0.3894	-0.0106
0.5	0.4794	-0.0206

(b) See answer to part (a) above.

(c)

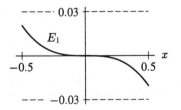

The fact that the graph of E_1 lies between the horizontal lines at ± 0.03 shows that $|E_1| < 0.03$ for $-0.5 \leq x \leq 0.5$.

CHAPTER TEN

Solutions for Section 10.1

1. (a) (III) An island can only sustain the population up to a certain size. The population will grow until it reaches this limiting value.
   (b) (V) The ingot will get hot and then cool off, so the temperature will increase and then decrease.
   (c) (I) The speed of the car is constant, and then decreases linearly when the breaks are applied uniformly.
   (d) (II) Carbon-14 decays exponentially.
   (e) (IV) Tree pollen is seasonal, and therefore cyclical.

5. If $y = \sin 2t$, then $\frac{dy}{dt} = 2\cos 2t$, and $\frac{d^2y}{dt^2} = -4\sin 2t$.
   Thus $\frac{d^2y}{dt^2} + 4y = -4\sin 2t + 4\sin 2t = 0$.

9. The easiest way to approach this problem is to take the derivative dy/dx of all functions (I)–(V), and try to match the formulas obtained with the equations (A)–(E).

   (I)
   $$y = x^3$$
   $$\frac{dy}{dx} = 3x^2 = 3\frac{x^3}{x} = 3\frac{y}{x},$$

   thus (I) is a solution to (B).

   (II)
   $$y = 3x$$
   $$\frac{dy}{dx} = 3 = \frac{y}{x},$$

   thus (II) is a solution to (A).

   (III)
   $$y = e^{3x}$$
   $$\frac{dy}{dx} = 3e^{3x} = 3y,$$

   thus (III) is a solution to (E).

   (IV)
   $$y = 3e^x$$
   $$\frac{dy}{dx} = 3e^x = y,$$

   thus (IV) is a solution to (D).

   (V)
   $$y = x$$
   $$\frac{dy}{dx} = 1 = \frac{y}{x},$$

   thus (V) is a solution to (A).

   We notice that none of the five functions is a solution to (C).

Solutions for Section 10.2

1. (a)

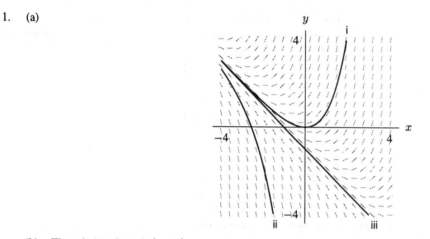

    (b)   The solution through $(-1, 0)$ appears to be linear, so its equation is $y = -x - 1$.
    (c)   If $y = -x - 1$, then $y' = -1$ and $x + y = x + (-x - 1) = -1$, so this checks as a solution.

5.   Notice that $y' = \dfrac{x + y}{x - y}$ is zero when $x = -y$ and is undefined when $x = y$. A solution curve will be horizontal (slope= 0) when passing through a point with $x = -y$, and will be vertical (slope undefined) when passing through a point with $x = y$. The only slope field for which this is true is slope field (b).

Solutions for Section 10.3

1. (a)

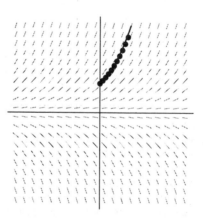

Figure 10.1

    (b) $y(0) = 1,$
    $$y(0.1) \approx y(0) + 0.1y(0) = 1 + 0.1(1) = 1.1$$
    $$y(0.2) \approx y(0.1) + 0.1y(0.1) = 1.1 + 0.1(1.1) = 1.21$$
    $$y(0.3) \approx y(0.2) + 0.1y(0.2) = 1.21 + 0.1(1.21) = 1.331$$
    $$y(0.4) \approx 1.4641$$
    $$y(0.5) \approx 1.61051$$
    $$y(0.6) \approx 1.77156$$
    $$y(0.7) \approx 1.94872$$
    $$y(0.8) \approx 2.14359$$
    $$y(0.9) \approx 2.35795$$
    $$y(1.0) \approx 2.59374$$
    (c)   See Figure 10.1. A smooth curve drawn through the solution points seems to match the slopefield.

(d) For $y = e^x$, we have $y' = e^x = y$ and $y(0) = e^0 = 1$.

	Computed Solution	
x_n	Approx. $y(x_n)$	$y(x_n)$
0	1	1
0.1	1.1	1.10517
0.2	1.21	1.22140
0.3	1.331	1.34986
0.4	1.4641	1.49182
0.5	1.61051	1.64872
0.6	1.77156	1.82212
0.7	1.94872	2.01375
0.8	2.14359	2.22554
0.9	2.35795	2.45960
1.0	2.59374	2.71828

5. (a)

TABLE 10.1

t	y	slope $= \frac{1}{t}$	$\Delta y = (\text{slope})\Delta t = \frac{1}{t}(0.1)$
1	0	1	0.1
1.1	0.1	0.909	0.091
1.2	0.191	0.833	0.083
1.3	0.274	0.769	0.077
1.4	0.351	0.714	0.071
1.5	0.422	0.667	0.067
1.6	0.489	0.625	0.063
1.7	0.552	0.588	0.059
1.8	0.610	0.556	0.056
1.9	0.666	0.526	0.053
2	0.719		

(b) If $\frac{dy}{dt} = \frac{1}{t}$, then $y = \ln|t| + C$.
Starting at $(1, 0)$ means $y = 0$ when $t = 1$, so $C = 0$ and $y = \ln|t|$.
After ten steps, $t = 2$, so $y = \ln 2 \approx 0.693$.

(c) Approximate $y = 0.719$, Exact $y = 0.693$.
Thus the approximate answer is too big. This is because the solution curve is concave down, and so the tangent lines are above the curve. Figure 10.2 shows the slope field of $y' = 1/t$ with the solution curve $y = \ln t$ plotted on top of it.

Figure 10.2

9.   (a)   Using one step, $\frac{\Delta B}{\Delta t} = 0.05$, so $\Delta B = \left(\frac{\Delta B}{\Delta t}\right)\Delta t = 50$. Therefore we get an approximation of $B \approx 1050$ after one year.
     (b)   With two steps, $\Delta t = 0.5$ and we have

TABLE 10.2

t	B	$\Delta B = (0.05B)\Delta t$
0	1000	25
0.5	1025	25.63
1.0	1050.63	

     (c)   Keeping track to the nearest hundredth with $\Delta t = 0.25$, we have

TABLE 10.3

t	B	$\Delta B = (0.05B)\Delta t$
0	1000	12.5
0.25	1012.5	12.66
0.5	1025.16	12.81
0.75	1037.97	12.97
1	1050.94	

     (d)   In part (a), we get our approximation by making a single increment, ΔB, where ΔB is just $0.05B$. If we think in terms of interest, ΔB is just like getting one end of the year interest payment. Since ΔB is 0.05 times the balance B, it is like getting 5% interest at the end of the year.
     (e)   Part (b) is equivalent to computing the final amount in an account that begins with $1000 and earns 5% interest compounded twice annually. Each step is like computing the interest after 6 months. When $t = 0.5$, for example, the interest is $\Delta B = (0.05B) \cdot \frac{1}{2}$, and we add this to $1000 to get the new balance.

     Similarly, part (c) is equivalent to the final amount in an account that has an initial balance of $1000 and earns 5% interest compounded quarterly.

Solutions for Section 10.4 ▬▬▬▬

1. $\frac{dP}{dt} = 0.02P$ implies that $\frac{dP}{P} = 0.02\,dt$.

     $\int \frac{dP}{P} = \int 0.02\,dt$ implies that $\ln|P| = 0.02t + C$.

     $|P| = e^{0.02t+C}$ implies that $P = Ae^{0.02t}$, where $A = \pm e^C$.
     We are given $P(0) = 20$. Therefore, $P(0) = Ae^{(0.02)\cdot 0} = A = 20$. So the solution is $P = 20e^{0.02t}$.

5. $\frac{dy}{dx} + \frac{y}{3} = 0$ implies $\frac{dy}{dx} = -\frac{y}{3}$ implies $\int \frac{dy}{y} = -\int \frac{1}{3}\,dx$.
     Integrating and moving terms, we have $y = Ae^{-\frac{1}{3}x}$. Since $y(0) = A = 10$, we have $y = 10e^{-\frac{1}{3}x}$.

9.   Factoring out the 0.1 gives $\frac{dm}{dt} = 0.1m + 200 = 0.1(m + 2000)$.
     $\frac{dm}{m+2000} = 0.1\,dt$ implies that $\int \frac{dm}{m+2000} = \int 0.1\,dt$, so $\ln|m + 2000| = 0.1t + C$. So $m = Ae^{0.1t} - 2000$. Using the initial condition, $m(0) = Ae^{(0.1)\cdot 0} - 2000 = 1000$, gives $A = 3000$. Thus $m = 3000e^{0.1t} - 2000$.

13. Rearrange and write

$$\int \frac{1}{1-R}\,dR = \int dr$$

or

$$-\ln|1-R| = r + C$$

which can be written as

$$1 - R = \pm e^{-C-r} = Ae^{-r}$$

or

$$R(r) = 1 - Ae^{-r}.$$

The initial condition $R(1) = 0.1$ gives $0.1 = 1 - Ae^{-1}$ and so

$$A = 0.9e.$$

Therefore

$$R(r) = 1 - 0.9e^{1-r}.$$

17. $\frac{dy}{dt} = y^2(1+t)$ implies that $\int \frac{dy}{y^2} = \int (1+t)\,dt$ implies that $-\frac{1}{y} = t + \frac{t^2}{2} + C$ implies that $y = -\frac{1}{t + \frac{t^2}{2} + C}$.
Since $y = 2$ when $t = 1$, then $2 = -\frac{1}{1 + \frac{1}{2} + C}$. So $2C + 3 = -1$, and $C = -2$. Thus $y = -\frac{1}{\frac{t^2}{2} + t - 2} = -\frac{2}{t^2 + 2t - 4}$.

21. Separating variables and integrating with respect to r gives

$$\int \frac{1}{z}\,dz = \int (1 + r^2)\,dr$$

so that

$$\ln|z| = r + \frac{1}{3}r^3 + C.$$

The initial condition $z(0) = 1$ gives $C = 0$ so that

$$z(r) = e^{r + (1/3)r^3}.$$

25. $\frac{dP}{dt} = P - a$, implying that $\frac{dP}{P-a} = dt$ so $\int \frac{dP}{P-a} = \int dt$. Integrating yields $\ln|P - a| = t + C$, so $|P - a| = e^{t+C} = e^t e^C$. $P = a + Ae^t$, where $A = \pm e^C$ or $A = 0$.

29. $\frac{dy}{dt} = y(2-y)$ which implies that $\frac{dy}{y(y-2)} = -dt$, implying that $\int \frac{dy}{(y-2)(y)} = -\int dt$, so $-\frac{1}{2} \int \left(\frac{1}{y} - \frac{1}{y-2}\right) dy = -\int dt$.
Integrating yields $\frac{1}{2}(\ln|y-2| - \ln|y|) = -t + C$, so $\ln \frac{|y-2|}{|y|} = -2t + 2C$.
Exponentiating both sides yields $|1 - \frac{2}{y}| = e^{-2t+2C} \Rightarrow \frac{2}{y} = 1 - Ae^{-2t}$, where $A = \pm e^{2C}$. Hence $y = \frac{2}{1 - Ae^{-2t}}$. But $y(0) = \frac{2}{1-A} = 1$, so $A = -1$, and $y = \frac{2}{1 + e^{-2t}}$.

33. (a), (b)

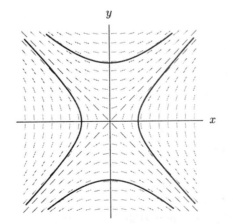

(c) $\frac{dy}{dx} = \frac{x}{y}$, so $\int y\,dy = \int x\,dx$ and thus $\frac{y^2}{2} = \frac{x^2}{2} + C$, or $y^2 - x^2 = 2C$. This is the equation of the hyperbolas in (b).

37.   Starting off with the homogeneous equation

$$f\left(\frac{y}{x}\right) = \frac{dy}{dx},$$

we replace $y(x)$ with $xv(x)$ and get

$$f(v) = f\left(\frac{xv}{x}\right) = f\left(\frac{y}{x}\right) = \frac{dy}{dx} = \frac{d(xv(x))}{dx} = v + x\frac{dv}{dx}.$$

Thus we have

$$f(v) - v = x\frac{dv}{dx}$$

which is clearly separable.

Solutions for Section 10.5

1.   (a)   If the world's population grows exponentially, satisfying $\frac{dP}{dt} = kP$, and if the arable land used is proportional to the population, then we'd expect A to satisfy $\frac{dA}{dt} = kA$. One is, of course, also assuming that the total amount of arable land is large compared to the amount that is now being used.

(b)   We have $A(t) = A_0 e^{kt} = (1 \times 10^9)e^{kt}$, where t is the number of years after 1950. Since $2 \times 10^9 = (1 \times 10^9)e^{k(30)}$, we have $e^{30k} = 2$, so $k = \frac{\ln 2}{30} \approx 0.023$. Thus, $A \approx (1 \times 10^9)e^{0.023t}$. We want to find t such that $3.2 \times 10^9 = A(t) = (1 \times 10^9)e^{0.023t}$. Taking logarithms yields

$$t = \frac{\ln(3.2)}{0.023} \approx 50.6 \text{ years}.$$

Thus the arable land will have run out by the year 2001.

5.   (a)   We know that the equilibrium solutions are the functions satisfying the differential equation whose derivative everywhere is 0. Thus we have

$$\frac{dy}{dt} = 0$$
$$0.2(y - 3)(y + 2) = 0$$
$$(y - 3)(y + 2) = 0.$$

The solutions are $y = 3$ and $y = -2$.

(b)

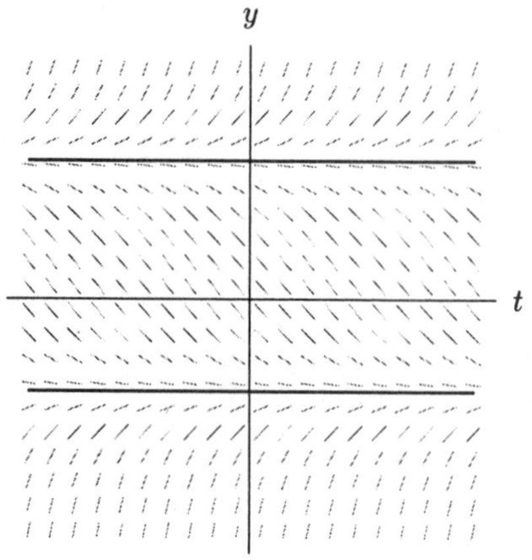

Figure 10.3

Looking at Figure 10.3, we see that the line $y = 3$ is an unstable solution, while the line $y = -2$ is a stable solution.

9. (a) $\frac{dB}{dt} = \frac{r}{100}B$. The constant of proportionality is $\frac{r}{100}$.

    (b) Solving, we have

$$\frac{dB}{B} = \frac{r\,dt}{100}$$

$$\int \frac{dB}{B} = \int \frac{r}{100}\,dt$$

$$\ln|B| = \frac{r}{100}t + C$$

$$B = e^{(r/100)t+C} = Ae^{(r/100)t}, \qquad A = e^{C}.$$

   A is the initial amount in the account, since A is the amount at time $t = 0$.

   (c)

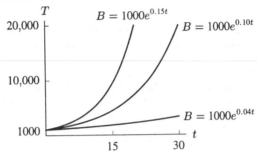

13. (a)

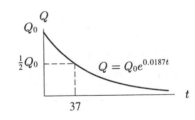

   (b) $\dfrac{dQ}{dt} = -kQ$

   (c) Since $Q = Q_0 e^{-kt}$ and $\frac{1}{2} = e^{-k(37)}$, we have

$$k = -\frac{\ln(1/2)}{37} = 0.0187.$$

   Therefore $Q = Q_0 e^{-0.0187t}$. We know that when the drug level is 25% of the original level that $Q = 0.25Q_0$. Setting these equal, we get

$$0.25Q_0 = e^{-0.0187t}.$$

   giving

$$t = -\frac{\ln(0.25)}{0.0187} \approx 74 \text{ hours} \approx 3 \text{ days}.$$

17. (a) If P = pressure and h = height, $\frac{dP}{dh} = -3.7 \times 10^{-5}P$, so $P = P_0 e^{-3.7\times10^{-5}h}$. Now $P_0 = 29.92$, since pressure at sea level (when $h = 0$) is 29.92, so $P = 29.92e^{-3.7\times10^{-5}h}$. At the top of Mt. Whitney, the pressure is

$$P = 29.92e^{-3.7\times10^{-5}(14500)} \approx 17.50 \text{ inches of mercury.}$$

   At the top of Mt. Everest, the pressure is

$$P = 29.92e^{-3.7\times10^{-5}(29000)} \approx 10.23 \text{ inches of mercury.}$$

   (b) The pressure is 15 inches of mercury when

$$15 = 29.92e^{-3.7\times10^{-5}h}$$

   Solving for h gives $h = \frac{-1}{3.7\times10^{-5}} \ln(\frac{15}{29.92}) \approx 18{,}661.5$ feet.

21. The rate of disintegration is proportional to the quantity of carbon-14 present. Let Q be the quantity of carbon-14 present at time t, with $t = 0$ in 1977. Then

$$Q = Q_0 e^{-kt},$$

where Q_0 is the quantity of carbon-14 present in 1977 when $t = 0$. Then we know that

$$\frac{Q_0}{2} = Q_0 e^{-k(5730)}$$

so that

$$k = -\frac{\ln(1/2)}{5730} = 0.000121.$$

Thus

$$Q = Q_0 e^{-0.000121t}.$$

The quantity present at any time is proportional to the rate of disintegration at that time so

$$Q_0 = c8.2 \qquad \text{and} \qquad Q = c13.5$$

where c is a constant of proportionality. Thus substituting for Q and Q_0 in

$$Q = Q_0 e^{-0.000121t}$$

gives

$$c13.5 = c8.2 e^{-0.000121t}$$

so

$$t = -\frac{\ln(13.5/8.2)}{0.000121} \approx -4120.$$

Thus Stonehenge was built about 4120 years before 1977, in about 2150 B.C.

Solutions for Section 10.6

1. Let $D(t)$ be the quantity of dead leaves, in grams per square centimeter. Then $\frac{dD}{dt} = 3 - 0.75D$, where t is in years. We factor out -0.75 and then separate variables.

$$\frac{dD}{dt} = -0.75(D - 4)$$
$$\int \frac{dD}{D - 4} = \int -0.75\,dt$$
$$\ln|D - 4| = -0.75t + C$$
$$|D - 4| = e^{-0.75t+C} = e^{-0.75t}e^{C}$$
$$D = 4 + Ae^{-0.75t}, \text{ where } A = \pm e^{C}.$$

If initially the ground is clear, the solution looks like:

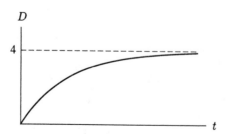

The equilibrium level is 4 grams per square centimeter, regardless of the initial condition.

5.   Let the depth of the water at time t be y. Then $\dfrac{dy}{dt} = -k\sqrt{y}$, where k is a positive constant. Separating variables,

$$\int \frac{dy}{\sqrt{y}} = -\int k\,dt,$$

so

$$2\sqrt{y} = -kt + C.$$

When $t = 0$, $y = 36$; $2\sqrt{36} = -k \cdot 0 + C$, so $C = 12$.
When $t = 1$, $y = 35$; $2\sqrt{35} = -k + 12$, so $k \approx 0.17$.
Thus, $2\sqrt{y} \approx -0.17t + 12$. We are looking for t such that $y = 0$; this happens when $t \approx \frac{12}{0.17} \approx 71$ hours, or about 3 days.

9.   (a)   Quantity of A present at time t equals $(a - x)$.
Quantity of B present at time t equals $(b - x)$.
So

$$\text{Rate of formation of } C = k(\text{Quantity of } A)(\text{Quantity of } B)$$

gives

$$\frac{dx}{dt} = k(a - x)(b - x)$$

(b)   Separating gives

$$\int \frac{dx}{(a - x)(b - x)} = \int k\,dt.$$

Rewriting the denominator as $(a - x)(b - x) = (x - a)(x - b)$ enables us to use Formula 26 in the Table of Integrals provided $a \neq b$. For some constant K, this gives

$$\frac{1}{a - b}\left(\ln|x - a| - \ln|x - b|\right) = kt + K.$$

Thus

$$\ln\left|\frac{x - a}{x - b}\right| = (a - b)kt + K(a - b)$$

$$\left|\frac{x - a}{x - b}\right| = e^{K(a-b)}e^{(a-b)kt}$$

$$\frac{x - a}{x - b} = Me^{(a-b)kt} \quad \text{where } M = \pm e^{K(a-b)}.$$

Since $x = 0$ when $t = 0$, we have $M = \frac{a}{b}$. Thus

$$\frac{x - a}{x - b} = \frac{a}{b}e^{(a-b)kt}.$$

Solving for x, we have

$$bx - ba = ae^{(a-b)kt}(x - b)$$

$$x(b - ae^{(a-b)kt}) = ab - abe^{(a-b)kt}$$

$$x = \frac{ab(1 - e^{(a-b)kt})}{b - ae^{(a-b)kt}} = \frac{ab(e^{bkt} - e^{akt})}{be^{bkt} - ae^{akt}}.$$

13.   (a) $\dfrac{dy}{dt} = -k(y - a)$, where $k > 0$ and a are constants.

(b) $\displaystyle\int \frac{dy}{y - a} = \int -k\,dt$, so $\ln|y - a| = \ln(y - a) = -kt + C$. Thus, $y - a = Ae^{-kt}$ where $A = e^C$. Initially nothing has been forgotten, so $y(0) = 1$. Therefore, $1 - a = Ae^0 = A$, so $y - a = (1 - a)e^{-kt}$ or $y = (1 - a)e^{-kt} + a$.

(c)   As $t \to \infty$, $e^{-kt} \to 0$, so $y \to a$.
Thus, a represents the fraction of material which is remembered in the long run. The constant k tells us about the rate at which material is forgotten.

17. (a) Now

$$\frac{dS}{dt} = (\text{Rate at which salt enters the pool}) - (\text{Rate at which salt leaves the pool}),$$

and, for example,

$$\left(\begin{array}{c}\text{Rate at which salt} \\ \text{enters the pool}\end{array}\right) = \left(\begin{array}{c}\text{Concentration of} \\ \text{salt solution}\end{array}\right) \times \left(\begin{array}{c}\text{Flow rate of} \\ \text{salt solution}\end{array}\right)$$

$$(\text{grams/minute}) = (\text{grams/liter}) \times (\text{liters/minute})$$

so

$$\text{Rate at which salt enters the pool} =$$

$$(10 \text{ grams/liter}) \times (60 \text{ liters/minute}) = (600 \text{ grams/minute})$$

The rate at which salt leaves the pool depends on the concentration of salt in the pool. At time t, the concentration is $\frac{S(t)}{2 \times 10^6 \text{ liters}}$, where $S(t)$ is measured in grams.
Thus

$$\text{Rate at which salt leaves the pool} =$$

$$\frac{S(t) \text{ grams}}{2 \times 10^6 \text{ liters}} \times \frac{60 \text{ liters}}{\text{minute}} = \frac{3S(t) \text{ grams}}{10^5 \text{ minutes}}.$$

Thus

$$\frac{dS}{dt} = 600 - \frac{3S}{100,000}.$$

(b) $\frac{dS}{dt} = -\frac{3}{100,000}(S - 20,000,000)$

$\int \frac{dS}{S - 20,000,000} = \int -\frac{3}{100,000} \, dt$

$\ln|S - 20,000,000| = -\frac{3}{100,000}t + C$

$S = 20,000,000 - Ae^{-\frac{3}{100,000}t}$

Since $S = 0$ at $t = 0$, $A = 20,000,000$. Thus $S(t) = 20,000,000 - 20,000,000e^{-\frac{3}{100,000}t}$.

(c) As $t \to \infty$, $e^{-\frac{3}{100,000}t} \to 0$, so $S(t) \to 20,000,000$ grams. The concentration approaches 10 grams/liter. Note that this makes sense; we'd expect the concentration of salt in the pool to become closer and closer to the concentration of salt being poured into the pool as $t \to \infty$.

21. (a) For a stable universe, we need $R' = 0$, so $R'' = 0$. However the differential equation for R'' shows that $R'' < 0$ for every R, so we never have $R'' = 0$. Thus R' and R must both be changing with time.

(b) If the universe were expanding at a constant rate of $R'(t_0)$, then $R(t_0)/R'(t_0)$ would be the time it took for the radius to grow from 0 to $R(t_0)$ – a reasonable estimate for the age of the universe. Since in fact R' has been decreasing, in other words, the universe has actually been expanding faster than $R'(t_0)$, the Hubble constant is an overestimate (i.e. the universe is actually younger than the Hubble constant suggests.)

Solutions for Section 10.7

1. A continuous growth rate of 0.2% means that

$$\frac{1}{P}\frac{dP}{dt} = 0.2\% = 0.002.$$

Separating variables and integrating gives

$$\int \frac{dP}{P} = \int 0.002 \, dt$$

$$P = P_0 e^{0.002t} = (6.6 \times 10^6)e^{0.002t}.$$

5. Rewriting the equation as $\frac{1}{P}\frac{dP}{dt} = \frac{(100-P)}{1000}$, we see that this is a logistic equation. Before looking at its solution, we explain why there must always be at least 100 individuals. Since the population begins at 200, $\frac{dP}{dt}$ is initially negative, so the population decreases. It continues to do so while $P > 100$. If the population ever reached 100, however, then $\frac{dP}{dt}$ would be 0. This means the population would stop changing – so if the population ever decreased to 100, that's where it would stay. The fact that $\frac{dP}{dt}$ will always be negative also shows that the population will always be under 200, as shown below.

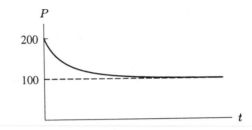

The solution, as given by the formula derived in the chapter, is

$$P = \frac{20000}{200 - 100e^{-t/10}}$$

9. (a) To estimate $\frac{dP}{dt}$ for 1810, for example, we'll take

$$\frac{(\text{pop. at 1820})-(\text{pop. at 1800})}{20 \text{ years}} = \frac{(9.6 - 5.3) \text{ million}}{20 \text{ years}} = 0.215\frac{\text{million}}{\text{yr}}.$$

Thus $\frac{1}{P}\frac{dP}{dt} = \frac{0.215}{7.2} \approx 0.03$. We do this for several other points.

Year	P	$\frac{dP}{dt} \approx \frac{P(t+10)-P(t-10)}{20}$	$\frac{1}{P}\frac{dP}{dt}$
1800	5.3	$(7.2 - 3.9)/20 = 0.165$	0.0311
1830	12.9	$(17.1 - 9.6)/20 = 0.375$	0.0291
1860	31.4	$(38.6 - 23.1)/20 = 0.775$	0.0247
1890	62.9	$(76.0 - 50.2)/20 = 1.29$	0.0205
1920	105.7	$(122.8 - 92.0)/20 = 1.54$	0.0146
1950	150.7	$(179.0 - 131.7)/20 = 2.365$	0.0157
1980	226.5	$(248.7 - 205.0)/20 = 2.185$	0.0096

Plotting the data and fitting a line to it, we obtain

$$\frac{1}{P}\frac{dP}{dt} = 0.0286 - 0.0001P.$$

Thus $k \approx 0.0286$ and $a \approx 0.0001$.

(b) According to this model, $\frac{1}{P}\frac{dP}{dt} = 0.0001(286 - P)$. Thus, P will increase up to about 286 million, and then level off.

13. Using a one-sided estimate for $f'(2)$, we get:

$$f'(2) \approx \frac{f(2+h) - f(2)}{h} = \frac{(2+h)^3 - (2)^3}{h}$$
$$= \frac{2^3 + 12h + 6h^2 + h^3 - 2^3}{h}$$
$$= 12 + 6h + h^2$$

If $h = 0.1$, we have $f'(2) \approx 12.61$.

If $h = 0.01$, we have $f'(2) \approx 12.0601$.

If $h = 0.001$, we have $f'(2) \approx 12.006001$.

As h decreases by a factor of ten, our approximation improves by one digit of accuracy.

Using a two-sided estimate for $f'(2)$, we get:

$$f'(2) \approx \frac{f(2+h) - f(2-h)}{2h} = \frac{(2+h)^3 - (2-h)^3}{2h}$$

$$= \frac{(2^3 + 12h + 6h^2 + h^3) - (2^3 - 12h + 6h^2 - h^3)}{2h}$$

$$= \frac{24h + 2h^3}{2h} = 12 + h^2$$

If $h = 0.1$, we have $f'(2) \approx 12.01$.

If $h = 0.01$, we have $f'(2) \approx 12.0001$.

If $h = 0.001$, we have $f'(2) \approx 12.000001$.

As h decreases by a factor of ten, the one-sided approximation improves by two digits of accuracy. The two-sided estimate is accurate to twice as many digits as the one-sided estimate.

17.

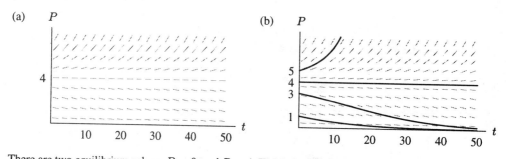

(a) P

(b) P

(c) There are two equilibrium values, $P = 0$, and $P = 4$. The first, representing extinction, is stable. The equilibrium value $P = 4$ is unstable because the populations increase if greater than 4, and decrease if less than 4. Notice that the equilibrium values can be obtained by setting $dP/dt = 0$:

$$\frac{dP}{dt} = 0.02P^2 - 0.08P = 0.02P(P - 4) = 0$$

so

$$P = 0 \text{ or } P = 4.$$

Solutions for Section 10.8

1. If $y = 2\cos t + 3\sin t$, then $y' = -2\sin t + 3\cos t$ and $y'' = -2\cos t - 3\sin t$. Thus, $y'' + y = 0$.

5. At $t = 0$, we find that $y = 0$. Since $-1 \le \sin 3t \le 1$, y ranges from -0.5 to 0.5, so at $t = 0$ it is starting in the middle. Since $y' = -1.5\cos 3t$, we see $y' = -1.5$ when $t = 0$, so the mass is moving downward.

9. First, we note that the solutions of:
(a) $x'' + x = 0$ are $x = A\cos t + B\sin t$;
(b) $x'' + 4x = 0$ are $x = A\cos 2t + B\sin 2t$;
(c) $x'' + 16x = 0$ are $x = A\cos 4t + B\sin 4t$.

This follows from what we know about the general solution to $x'' + \omega^2 x = 0$.

The period of the solutions to (a) is 2π, the period of the solutions to (b) is π, and the period of the solutions of (c) is $\frac{\pi}{2}$. Since the t-scales are the same on all of the graphs, we see that graphs (I) and (IV) have the same period, which is twice the period of graph (III). Graph (II) has twice the period of graphs (I) and (IV). Since each graph represents a solution, we have the following:

- equation (a) goes with graph (II)
  equation (b) goes with graphs (I) and (IV)
  equation (c) goes with graph (III)
- The graph of (I) passes through $(0,0)$, so $0 = A\cos 0 + B\sin 0 = A$. Thus, the equation is $x = B\sin 2t$. Since the amplitude is 2, we see that $x = 2\sin 2t$ is the equation of the graph. Similarly, the equation for (IV) is $x = -3\sin 2t$.

  The graph of (II) also passes through $(0,0)$, so, similarly, the equation must be $x = B\sin t$. In this case, we see that $B = -1$, so $x = -\sin t$.

  Finally, the graph of (III) passes through $(0,1)$, and 1 is the maximum value. Thus, $1 = A\cos 0 + B\sin 0$, so $A = 1$. Since it reaches a local maximum at $(0,1)$, $x'(0) = 0 = -4A\sin 0 + 4B\cos 0$, so $B = 0$. Thus, the solution is $x = \cos 4t$.

13. The amplitude is $\sqrt{3^2 + 7^2} = \sqrt{58}$.

17. **(a)** Let $x = \omega t$ and $y = \phi$. Then

    $$A\sin(\omega t + \phi) = A(\sin\omega t\cos\phi + \cos\omega t\sin\phi)$$
    $$= (A\sin\phi)\cos\omega t + (A\cos\phi)\sin\omega t.$$

    (b) If we want $A\sin(\omega t + \phi) = C_1\cos\omega t + C_2\sin\omega t$ to be true for all t, then by looking at the answer to part (a), we must have $C_1 = A\sin\phi$ and $C_2 = A\cos\phi$. Thus,

    $$\frac{C_1}{C_2} = \frac{A\sin\phi}{A\cos\phi} = \tan\phi,$$

    and

    $$\sqrt{C_1^2 + C_2^2} = \sqrt{A^2\sin^2\phi + A^2\cos^2\phi} = A\sqrt{\sin^2\phi + \cos^2\phi} = A,$$

    so our formulas are justified.

21. The equation we have for the charge tells us that:

    $$\frac{d^2Q}{dt^2} = -\frac{Q}{LC},$$

    where L and C are positive.

    If we let $\omega = \sqrt{\frac{1}{LC}}$, we know the solution is of the form:

    $$Q = C_1\cos\omega t + C_2\sin\omega t.$$

    Since $Q(0) = 0$, we find that $C_1 = 0$, so $Q = C_2\sin\omega t$.

    Since $Q'(0) = 4$, and $Q' = \omega C_2\cos\omega t$, we have $C_2 = \frac{4}{\omega}$, so $Q = \frac{4}{\omega}\sin\omega t$.

    But we want the maximum charge, meaning the amplitude of Q, to be $2\sqrt{2}$ coulombs. Thus, we have $\frac{4}{\omega} = 2\sqrt{2}$, which gives us $\omega = \sqrt{2}$.

    So we now have: $\sqrt{2} = \frac{1}{\sqrt{LC}} = \frac{1}{\sqrt{10C}}$. Thus, $C = \frac{1}{20}$ farads.

Solutions for Section 10.9

1. The characteristic equation is $r^2 + 4r + 3 = 0$, so $r = -1$ or -3.
   Therefore $y(t) = C_1 e^{-t} + C_2 e^{-3t}$.

5. The characteristic equation is $r^2 + 7 = 0$, so $r = \pm\sqrt{7}i$.
   Therefore $s(t) = C_1\cos\sqrt{7}t + C_2\sin\sqrt{7}t$.

9. The characteristic equation is $r^2 + r + 1 = 0$, so $r = -\frac{1}{2} \pm \frac{\sqrt{3}}{2}i$.
   Therefore $p(t) = C_1 e^{-t/2}\cos\frac{\sqrt{3}}{2}t + C_2 e^{-t/2}\sin\frac{\sqrt{3}}{2}t$.

13. The characteristic equation is $r^2 + 6r + 5 = 0$, so $r = -1$ or -5.
 Therefore $y(t) = C_1 e^{-t} + C_2 e^{-5t}$.
 $y'(t) = -C_1 e^{-t} - 5C_2 e^{-5t}$
 $y'(0) = 0 = -C_1 - 5C_2$
 $y(0) = 1 = C_1 + C_2$
 Therefore $C_2 = -1/4$, $C_1 = 5/4$ and $y(t) = \frac{5}{4}e^{-t} - \frac{1}{4}e^{-5t}$.

17. The characteristic equation is $r^2 + 2r + 2 = 0$, so $r = -1 \pm i$.
 Therefore $p(t) = C_1 e^{-t}\cos t + C_2 e^{-t}\sin t$.
 $p(0) = 0 = C_1$ so $p(t) = C_2 e^{-t}\sin t$
 $p(\pi/2) = 20 = C_2 e^{-\pi/2}\sin\frac{\pi}{2}$ so $C_2 = 20e^{\pi/2}$
 Therefore $p(t) = 20e^{\frac{\pi}{2}}e^{-t}\sin t = 20e^{\frac{\pi}{2}-t}\sin t$.

21. $0 = \frac{d^2}{dt^2}(e^{2t}) - 5\frac{d}{dt}(e^{2t}) + ke^{2t} = 4e^{2t} - 10e^{2t} + ke^{2t} = e^{2t}(k-6)$. Since $e^{2t} \neq 0$, we must have $k - 6 = 0$. Therefore $k = 6$.

 The characteristic equation is $r^2 - 5r + 6 = 0$, so $r = 2$ or 3. Therefore $y(t) = C_1 e^{2t} + C_2 e^{3t}$.

25. The frictional force is $F_{\text{drag}} = -c\frac{ds}{dt}$. Thus spring (iv) has the smallest frictional force.

29. Recall that $s'' + bs' + cs = 0$ is overdamped if the discriminant $b^2 - 4c > 0$, critically damped if $b^2 - 4c = 0$, and underdamped if $b^2 - 4c < 0$. Since $b^2 - 4c = 8 - 4c$, the solution is overdamped if $c < 2$, critically damped if $c = 2$, and underdamped if $c > 2$.

33. The differential equation is $Q'' + 2Q' + \frac{1}{4}Q = 0$, so the characteristic equation is $r^2 + 2r + \frac{1}{4} = 0$. This has roots $\frac{-2 \pm \sqrt{3}}{2} = -1 \pm \frac{\sqrt{3}}{2}$. Thus, the general solution is

$$Q(t) = C_1 e^{(-1+\frac{\sqrt{3}}{2})t} + C_2 e^{(-1-\frac{\sqrt{3}}{2})t},$$
$$Q'(t) = C_1\left(-1 + \frac{\sqrt{3}}{2}\right)e^{(-1+\frac{\sqrt{3}}{2})t} + C_2\left(-1 - \frac{\sqrt{3}}{2}\right)e^{(-1-\frac{\sqrt{3}}{2})t}.$$

We have

(a)

$$Q(0) = C_1 + C_2 = 0$$
$$\text{and} \quad Q'(0) = \left(-1 + \frac{\sqrt{3}}{2}\right)C_1 + \left(-1 - \frac{\sqrt{3}}{2}\right)C_2 = 2.$$

Using the formula for $Q(t)$, we have $C_1 = -C_2$. Using the formula for $Q'(t)$, we have:

$$2 = \left(-1 + \frac{\sqrt{3}}{2}\right)(-C_2) + \left(-1 - \frac{\sqrt{3}}{2}\right)C_2 = -\sqrt{3}C_2$$
$$\text{so,} \quad C_2 = -\frac{2}{\sqrt{3}}.$$

Thus, $C_1 = \frac{2}{\sqrt{3}}$, and $Q(t) = \frac{2}{\sqrt{3}}\left(e^{(-1+\frac{\sqrt{3}}{2})t} - e^{(-1-\frac{\sqrt{3}}{2})t}\right)$.

(b) We have

$$Q(0) = C_1 + C_2 = 2$$
$$\text{and} \quad Q'(0) = \left(-1 + \frac{\sqrt{3}}{2}\right)C_1 + \left(-1 - \frac{\sqrt{3}}{2}\right)C_2 = 0.$$

Using the first equation, we have $C_1 = 2 - C_2$. Thus,

$$\left(-1 + \frac{\sqrt{3}}{2}\right)(2 - C_2) + \left(-1 - \frac{\sqrt{3}}{2}\right)C_2 = 0$$

$$-\sqrt{3}C_2 = 2 - \sqrt{3}$$

$$C_2 = -\frac{2 - \sqrt{3}}{\sqrt{3}}$$

and $$C_1 = 2 - C_2 = \frac{2 + \sqrt{3}}{\sqrt{3}}.$$

Thus, $Q(t) = \frac{1}{\sqrt{3}}\left((2 + \sqrt{3})e^{(-1 + \frac{\sqrt{3}}{2})t} - (2 - \sqrt{3})e^{(-1 - \frac{\sqrt{3}}{2})t}\right).$

37.   In the overdamped case, we have a solution of the form

$$s = C_1 e^{r_1 t} + C_2 e^{r_2 t}$$

where r_1 and r_2 are real. We find a t such that $s = 0$, hence $C_1 e^{r_1 t} = -C_2 e^{r_2 t}$.

If $C_2 = 0$, then $C_1 = 0$, hence $s = 0$ for all t. But this doesn't match with Figure 37, so $C_2 \neq 0$. We divide by $C_2 e^{r_1 t}$, and get:

$$-\frac{C_1}{C_2} = e^{(r_2 - r_1)t}, \quad \text{where} -\frac{C_1}{C_2} > 0,$$

so the exponential is always positive. Therefore

$$(r_2 - r_1)t = \ln(-\frac{C_1}{C_2})$$

and

$$t = \frac{\ln(-\frac{C_1}{C_2})}{(r_2 - r_1)}.$$

So the mass passes through the equilibrium point only once, when $t = \frac{\ln(-\frac{C_1}{C_2})}{(r_2 - r_1)}$.

Solutions for Chapter 10 Review

1.   Using the solution of the logistic equation given on page 538 in Section 10.7, and using $y(0) = 1$, we get $y = \frac{10}{1 + 9e^{-10t}}$.

5. $\frac{df}{dx} = \sqrt{xf(x)}$ gives $\int \frac{df}{\sqrt{f(x)}} = \int \sqrt{x}\,dx$, so $2\sqrt{f(x)} = \frac{2}{3}x^{\frac{3}{2}} + C$. Since $f(1) = 1$, we have $2 = \frac{2}{3} + C$ so $C = \frac{4}{3}$.
Thus, $2\sqrt{f(x)} = \frac{2}{3}x^{\frac{3}{2}} + \frac{4}{3}$, so $f(x) = (\frac{1}{3}x^{\frac{3}{2}} + \frac{2}{3})^2$.
(Note: this is only defined for $x \geq 0$.)

9. $\frac{dy}{dx} + xy^2 = 0$ means $\frac{dy}{dx} = -xy^2$, so $\int \frac{dy}{y^2} = \int -x\,dx$ giving $-\frac{1}{y} = -\frac{x^2}{2} + C$. Since $y(1) = 1$ we have $-1 = -\frac{1}{2} + C$
so $C = -\frac{1}{2}$. Thus, $-\frac{1}{y} = -\frac{x^2}{2} - \frac{1}{2}$ giving $y = \frac{2}{x^2 + 1}$.

13. $\frac{dy}{dt} = 2^y \sin^3 t$ implies $\int 2^{-y}\,dy = \int \sin^3 t\,dt$. Using Integral Table Formula 17, we have

$$-\frac{1}{\ln 2}2^{-y} = -\frac{1}{3}\sin^2 t \cos t - \frac{2}{3}\cos t + C.$$

According to the initial conditions: $y(0) = 0$ so $-\frac{1}{\ln 2} = -\frac{2}{3} + C$, and $C = \frac{2}{3} - \frac{1}{\ln 2}$. Thus,

$$-\frac{1}{\ln 2}2^{-y} = -\frac{1}{3}\sin^2 t \cos t - \frac{2}{3}\cos t + \frac{2}{3} - \frac{1}{\ln 2}.$$

Solving for y gives:

$$2^{-y} = \frac{\ln 2}{3}\sin^2 t \cos t + \frac{2\ln 2}{3}\cos t - \frac{2\ln 2}{3} + 1.$$

Taking natural logs, (Notice the right side is always > 0.)

$$y \ln 2 = -\ln\left(\frac{\ln 2}{3}\sin^2 t \cos t + \frac{2\ln 2}{3}\cos t - \frac{2\ln 2}{3} + 1\right),$$

so

$$y = \frac{-\ln\left(\frac{\ln 2}{3}\sin^2 t \cos t + \frac{2\ln 2}{3}\cos t - \frac{2\ln 2}{3} + 1\right)}{\ln 2}$$

17. $\frac{dQ}{dt} = -t^2Q^2 - Q^2 + 4t^2 + 4 = -Q^2(t^2+1) + 4(t^2+1) = (t^2+1)(4-Q^2)$. Separating variables yields $\frac{dQ}{4-Q^2} = (t^2+1)\,dt$, so

$$-\int \frac{dQ}{(Q-2)(Q+2)} = -\frac{1}{4}\int \left(\frac{1}{Q-2} - \frac{1}{Q+2}\right)dQ = \int (t^2+1)\,dt.$$

Integrating, we obtain $-\frac{1}{4}(\ln|Q-2| - \ln|Q+2|) = \frac{t^3}{3} + t + C$, so $\ln\frac{|Q-2|}{|Q+2|} = -\frac{4t^3}{3} - 4t - 4C$. Exponentiating yields $\left|\frac{Q-2}{Q+2}\right| = e^{-\frac{4t^3}{3}-4t}e^{-4C}$. $\frac{Q-2}{Q+2} = Ae^{-\frac{4t^3}{3}-4t}$ where $A = \pm e^{4C}$. Solving for Q, $Q = \dfrac{4}{1 - Ae^{-\frac{4t^3}{3}-4t}} - 2$. Notice that A could be any constant, including 0. In fact, we also lost the solution $Q = -2$ when we divided both sides by $4 - Q^2$. (The solution $Q = 2$ corresponds to $A = 0$, but $Q = -2$, another valid solution, is lost by our division.)

21. (a) $\Delta x = \frac{1}{5} = 0.2$.

At $x = 0$:

$y_0 = 1, y' = 4$; so $\Delta y = 4(0.2) = 0.8$. Thus, $y_1 = 1 + 0.8 = 1.8$.

At $x = 0.2$:

$y_1 = 1.8, y' = 3.2$; so $\Delta y = 3.2(0.2) = 0.64$. Thus, $y_2 = 1.8 + 0.64 = 2.44$.

At $x = 0.4$:

$y_2 = 2.44, y' = 2.56$; so $\Delta y = 2.56(0.2) = 0.512$. Thus, $y_3 = 2.44 + 0.512 = 2.952$.

At $x = 0.6$:

$y_3 = 2.952, y' = 2.048$; so $\Delta y = 2.048(0.2) = 0.4096$. Thus, $y_4 = 3.3616$.

At $x = 0.8$:

$y_4 = 3.3616, y' = 1.6384$; so $\Delta y = 1.6384(0.2) = 0.32768$. Thus, $y_5 = 3.68928$. So $y(1) \approx 3.689$.

(b)

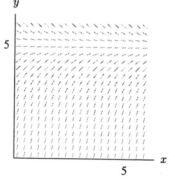

Since solution curves are concave down for $0 \le y \le 5$, and $y(0) = 1 < 5$, the estimate from Euler's method will be an overestimate.

(c) Solving by separation:

$$\int \frac{dy}{5-y} = \int dx, \quad \text{so} \quad -\ln|5-y| = x + C.$$

Then $5 - y = Ae^{-x}$ where $A = \pm e^{-C}$. Since $y(0) = 1$, we have $5 - 1 = Ae^0$, so $A = 4$.

Therefore, $y = 5 - 4e^{-x}$, and $y(1) = 5 - 4e^{-1} \approx 3.528$.

(Note: as predicted, the estimate in (a) is too large.)

(d) Doubling the value of n will probably halve the error and, therefore, give a value half way between 3.528 and 3.689, which is approximately 3.61.

25. The characteristic equation of $9z'' - z = 0$ is

$$9r^2 - 1 = 0.$$

If this is written in the form $r^2 + br + c = 0$, we have that $r^2 - 1/9 = 0$ and

$$b^2 - 4c = 0 - (4)(-1/9) = 4/9 > 0$$

This indicates overdamped motion and since the roots of the characteristic equation are $r = \pm 1/3$, the general solution is

$$y(t) = C_1 e^{\frac{1}{3}t} + C_2 e^{-\frac{1}{3}t}.$$

29. (a) Since the amount leaving the blood is proportional to the quantity in the blood,

$$\frac{dQ}{dt} = -kQ \quad \text{for some positive constant } k.$$

Thus $Q = Q_0 e^{-kt}$, where Q_0 is the initial quantity in the bloodstream. Only 20% is left in the blood after 3 hours. Thus $0.20 = e^{-3k}$, so $k = \frac{\ln 0.20}{-3} \approx 0.5365$. Therefore $Q = Q_0 e^{-0.5365t}$.

(b) Since 20% is left after 3 hours, after 6 hours only 20% of that 20% will be left. Thus after 6 hours only 4% will be left, so if the patient is given 100 mg, only 4 mg will be left 6 hours later.

33. (a) For this situation,

$$\begin{pmatrix} \text{Rate money added} \\ \text{to account} \end{pmatrix} = \begin{pmatrix} \text{Rate money added} \\ \text{via interest} \end{pmatrix} + \begin{pmatrix} \text{Rate money} \\ \text{deposited} \end{pmatrix}$$

Translating this into an equation yields

$$\frac{dB}{dt} = 0.1B + 1200$$

(b) Solving this equation via separation of variables gives

$$\frac{dB}{dt} = 0.1B + 1200$$
$$= (0.1)(B + 12000)$$

So

$$\int \frac{dB}{B + 12000} = \int 0.1\, dt$$

and

$$\ln|B + 12000| = 0.1t + C$$

solving for B,

$$|B + 12000| = e^{(0.1)t + C} = e^C e^{(0.1)t}$$

or

$$B = Ae^{0.1t} - 12000, \quad (\text{where } A = e^c)$$

We may find A using the initial condition $B_0 = f(0) = 0$

$$A - 12000 = 0 \quad \text{or} \quad A = 12000$$

So the solution is

$$B = f(t) = 12000(e^{0.1t} - 1)$$

(c) After 5 years, the balance is

$$B = f(5) = 12000(e^{(0.1)(5)} - 1)$$
$$= 7784.66$$

37. (a) When Juliet loves Romeo (i.e. $j > 0$), Romeo's love for her decreases (i.e. $\frac{dr}{dt} < 0$). When Juliet hates Romeo ($j < 0$), Romeo's love for her grows ($\frac{dr}{dt} > 0$). So j and $\frac{dr}{dt}$ have opposite signs, corresponding to the fact that $-B < 0$. When Romeo loves Juliet ($r > 0$), Juliet's love for him grows ($\frac{dj}{dt} > 0$). When Romeo hates Juliet ($r < 0$), Juliet's love for him decreases ($\frac{dj}{dt} < 0$). Thus r and $\frac{dj}{dt}$ have the same sign, corresponding to the fact that $A > 0$.

(b) Since $\frac{dr}{dt} = -Bj$, we have

$$\frac{d^2r}{dt^2} = \frac{d}{dt}(-Bj) = -B\frac{dj}{dt} = -ABr.$$

Rewriting the above equation as $r'' + ABr = 0$, we see that the characteristic equation is $R^2 + AB = 0$. Therefore $R = \pm\sqrt{AB}i$ and the general solution is

$$r(t) = C_1 \cos\sqrt{AB}t + C_2 \sin\sqrt{AB}t.$$

(c) Using $\frac{dr}{dt} = -Bj$, and differentiating r to find j, we obtain

$$j(t) = -\frac{1}{B}\frac{dr}{dt} = -\frac{\sqrt{AB}}{B}(-C_1 \sin \sqrt{AB}t + C_2 \cos \sqrt{AB}t).$$

Now, $j(0) = 0$ gives $C_2 = 0$ and $r(0) = 1$ gives $C_1 = 1$. Therefore, the particular solutions are

$$r(t) = \cos \sqrt{AB}t \quad \text{and} \quad j(t) = \sqrt{\frac{A}{B}} \sin \sqrt{AB}t$$

(d) Consider one period of the graph of $j(t)$ and $r(t)$:

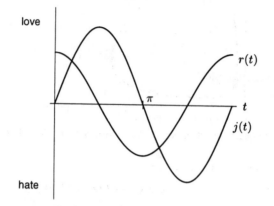

From the graph, we see that they both love each other only a quarter of the time.

Solutions to Problems on Systems of Differential Equations

1. Since

$$\frac{dS}{dt} = -aSI,$$

$$\frac{dI}{dt} = aSI - bI,$$

$$\frac{dR}{dt} = bI$$

we have

$$\frac{dS}{dt} + \frac{dI}{dt} + \frac{dR}{dt} = -aSI + aSI - bI + bI = 0.$$

Thus $\frac{d}{dt}(S + I + R) = 0$, so $S + I + R = \text{constant}$.

5. The closed trajectory represents populations which oscillate repeatedly.

9. If $w = 2$ and $r = 2$, then $\frac{dw}{dt} = -2$ and $\frac{dr}{dt} = 2$, so initially the number of worms decreases and the number of robins increases. In the long run, however, the populations will oscillate; they will even go back to $w = 2$ and $r = 2$.

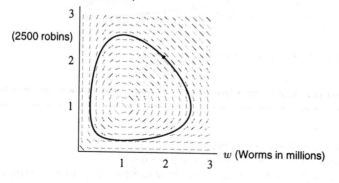

13. The numbers of robins begins to increase while the number of worms remains approximately constant.

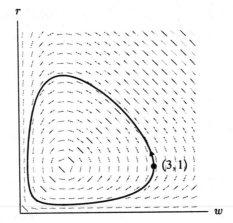

The numbers of robins and worms oscillate periodically between 0.2 and 3, with the robin population lagging behind the worm population.

17. (a) Lanchester's square law for the battle of Iwo Jima is

$$0.05y^2 - 0.01x^2 = C.$$

If we measure x and y in thousands, $x_0 = 54$ and $y_0 = 21.5$, so $0.05(21.5)^2 - 0.01(54)^2 = C$ giving $C = -6.0475$. Thus the equation of the trajectory is

$$0.05y^2 - 0.01x^2 = -6.0475$$

giving

$$x^2 - 5y^2 = 604.75.$$

(b) Assuming that the battle did not end until all the Japanese were dead or wounded, that is, $y = 0$, then the number of US soldiers remaining is given by $x^2 - 5(0)^2 = 604.75$. This gives $x = 24.59$, or about 25,000 troops. This is approximately what happened.

21. (a) Symbiosis, because both populations decrease while alone but are helped by the presence of the other.

(b)

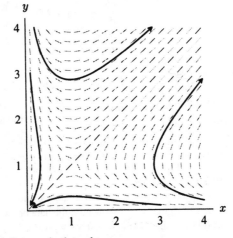

Both populations tend to infinity or both tend to zero.

Solutions to Problems on Analyzing the Phase Plane

1. (a) $dS/dt = 0$ where $S = 0$ or $I = 0$ (both axes).
    $dI/dt = 0.0026I(S - 192)$, so $dI/dt = 0$ where $I = 0$ or $S = 192$.
    Thus every point on the S axis is an equilibrium point (corresponding to no one being sick).

(b) In region I, where $S > 192$, $\dfrac{dS}{dt} < 0$ and $\dfrac{dI}{dt} > 0$.

In region II, where $S < 192$, $\dfrac{dS}{dt} < 0$ and $\dfrac{dI}{dt} < 0$. See Figure 10.4.

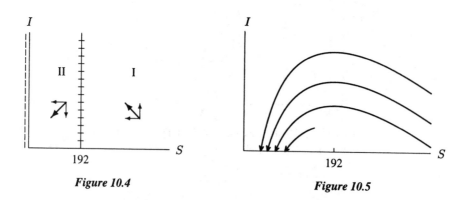

Figure 10.4   *Figure 10.5*

(c) If the trajectory starts with $S_0 > 192$, then I increases to a maximum when $S = 192$. If $S_0 < 192$, then I always decreases. See Figure 10.4. Regardless of the initial conditions, the trajectory always goes to a point on the S-axis (where $I = 0$). The S-intercept represents the number of students who never get the disease. See Figure 10.5.

5. We first find nullclines. Vertical nullclines occur where $\frac{dx}{dt} = x(2 - x - 2y) = 0$, which happens when $x = 0$ or $y = \frac{1}{2}(2 - x)$. Horizontal nullclines occur where $\frac{dy}{dt} = y(1 - 2x - y) = 0$, which happens when $y = 0$ or $y = 1 - 2x$. These nullclines are shown in Figure 10.6.

Equilibrium points (also shown in the figure below) occur at the intersections of vertical and horizontal nullclines. There are three such points for this system; $(0, 0)$, $(0, 1)$, and $(2, 0)$.

The nullclines divide the positive quadrant into three regions as shown in the figure below. Trajectory directions for these regions are shown in Figure 10.7.

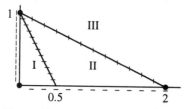

Figure 10.6: Nullclines and equilibrium
points (dots)

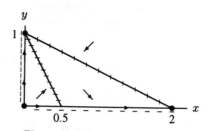

Figure 10.7: General directions of
trajectories and equilibrium points (dots)

9. (a) The nullclines are $P = 0$ or $P_1 + 3P_2 = 13$ (where $dP_1/dt = 0$) and $P = 0$ or $P_2 + 0.4P_1 = 6$ (where $dP_2/dt = 0$).

(b) The phase plane in Figure 10.8 shows that P_2 will eventually exclude P_1 regardless of where the experiment starts so long as there were some P_2 originally. Consequently, the data points would have followed a trajectory that starts at the origin, crosses the first nullcline and goes left and upwards between the two nullclines to the point $P_1 = 0$, $P_2 = 6$.

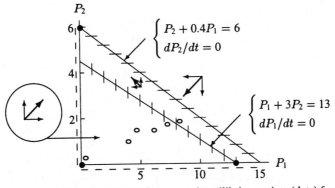

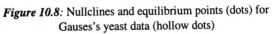

Figure 10.8: Nullclines and equilibrium points (dots) for Gauses's yeast data (hollow dots)

APPENDIX

Solutions for Section A

1. The graph is

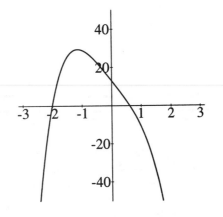

   (a) The range appears to be $y \leq 30$.
   (b) The function has two zeros.

5. The largest root is at about 2.5.

9. Using a graphing calculator, we see that when x is around 0.45, the graphs intersect.

13. (a) Only one real zero, at about $x = -1.15$.
   (b) Three real zeros: at $x = 1$, and at about $x = 1.41$ and $x = -1.41$.

17. (a) Since f is continuous, there must be one zero between $\theta = 1.4$ and $\theta = 1.6$, and another between $\theta = 1.6$ and $\theta = 1.8$. These are the only clear cases. We might also want to investigate the interval $0.6 \leq \theta \leq 0.8$ since $f(\theta)$ takes on values close to zero on at least part of this interval. Now, $\theta = 0.7$ is in this interval, and $f(0.7) = -0.01 < 0$, so f changes sign twice between $\theta = 0.6$ and $\theta = 0.8$ and hence has two zeros on this interval (assuming f is not *really* wiggly here, which it's not). There are a total of 4 zeros.
   (b) As an example, we find the zero of f between $\theta = 0.6$ and $\theta = 0.7$. $f(0.65)$ is positive; $f(0.66)$ is negative. So this zero is contained in $[0.65, 0.66]$. The other zeros are contained in the intervals $[0.72, 0.73]$, $[1.43, 1.44]$, and $[1.7, 1.71]$.
   (c) You've found all the zeros. A picture will confirm this; see Figure A.1.

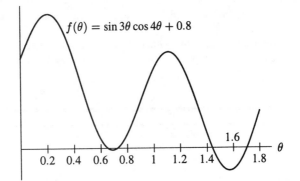

Figure A.1

21.

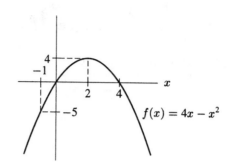

Bounded and $-5 \leq f(x) \leq 4$.

Solutions for Section B

1. For $20 \leq x \leq 100$, $0 \leq y \leq 1.2$, this function looks like a horizontal line at $y \approx 1.0725$ (In fact, the graph approaches this line from below.) Now, $e^{0.07} \approx 1.0725$, which strongly suggests that, as we already know, as $x \to \infty$, $\left(1 + \frac{0.07}{x}\right)^x \to e^{0.07}$.

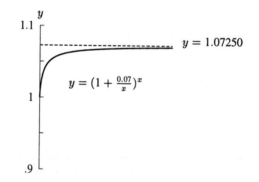

5. $e^{0.06} = 1.0618365$, so the effective annual rate $\approx 6.18365\%$.

9. (a) Using the formula $A = A_0(1 + \frac{r}{n})^{nt}$, we have $A = 10^6(1 + \frac{1}{12})^{12} \approx 10^6(2.61303529) \approx 2,613,035$ zaïre after one year.

    (b) Compounding daily, $A = 10^6(1 + \frac{1}{365})^{365} \approx 10^6(2.714567) \approx 2,714,567$ zaïre. Compounding hourly, $A = 10^6(1 + \frac{1}{8760})^{8760} \approx 10^6(2.7181267) \approx 2,718,127$ zaïre. Compounding each minute, $A = 10^6(1 + \frac{1}{525600})^{525600} \approx 10^6(2.718280) \approx 2,718,280$ zaïre

    (c) The amount does not seem to be increasing without bound, but rather it seems to level off at a value just over 2,718,000 zaïre. A close upper limit might be 2,718,300 (amounts may vary). In fact, the limit is $(e \times 10^6)$ zaïre.

Solutions for Section C

1. (1,0)

5. $(\frac{5\sqrt{3}}{2}, -\frac{5}{2})$

9. $r = \sqrt{0^2 + 2^2} = 2$, $\theta = \pi/2$.

13. $r = \sqrt{(0.2)^2 + (-0.2)^2} = 0.28$.

$\tan\theta = 0.2/(-0.2) = -1$. Since the point is in the fourth quadrant, $\theta = 7\pi/4$. (Alternatively $\theta = -\pi/4$.)

17. Putting $\theta = \pi/3$ into $\tan\theta = y/x$ gives $\sqrt{3} = y/x$, or $y = \sqrt{3}x$. This is a line through the origin of slope $\sqrt{3}$.

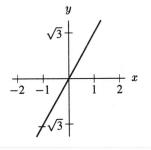

21. We multiply $r = \sin\theta/\cos^2\theta$ by $r\cos^2\theta$ and get $r^2\cos^2\theta = r\sin\theta$. Since $r\cos\theta = x$ and $r\sin\theta = y$, we have $y = x^2$, a parabola.

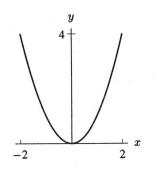

Solutions for Section D

1. $2e^{\frac{i\pi}{2}}$

5. $0e^{i\theta}$, for any θ.

9. $-3 - 4i$

13. $\frac{1}{4} - \frac{9i}{8}$

17. $5^3(\cos\frac{3\pi}{2} + i\sin\frac{3\pi}{2}) = -125i$

21. One value of $\sqrt[3]{i}$ is $\sqrt[3]{e^{i\frac{\pi}{2}}} = (e^{i\frac{\pi}{2}})^{\frac{1}{3}} = e^{i\frac{\pi}{6}} = \cos\frac{\pi}{6} + i\sin\frac{\pi}{6} = \frac{\sqrt{3}}{2} + \frac{i}{2}$

25. One value of $(-4 + 4i)^{2/3}$ is $[\sqrt{32}e^{i\frac{3\pi}{4}}]^{2/3} = (\sqrt{32})^{2/3}e^{i\frac{\pi}{2}} = 2^{\frac{10}{3}}\cos\frac{\pi}{2} + i2^{\frac{10}{3}}\sin\frac{\pi}{2} = 8i\sqrt[3]{2}$

29. Substituting $A_1 = 2 - A_2$ into the second equation gives

$$(1 - i)(2 - A_2) + (1 + i)A_2 = 0$$

so

$$2iA_2 = -2(1 - i)$$
$$A_2 = \frac{-(1 - i)}{i} = \frac{-i(1 - i)}{i^2} = i(1 - i) = 1 + i$$

Therefore $A_1 = 2 - (1 + i) = 1 - i$.

33. True, since $\sqrt{a}$ is real for all $a \geq 0$.

37. True. We can write any nonzero complex number z as $re^{i\beta}$, where r and β are real numbers with $r > 0$. Since $r > 0$, we can write $r = e^c$ for some real number c. Therefore, $z = re^{i\beta} = e^c e^{i\beta} = e^{c+i\beta} = e^w$ where $w = c + i\beta$ is a complex number.

41. Using Euler's formula, we have:

$$e^{i(2\theta)} = \cos 2\theta + i \sin 2\theta$$

On the other hand,

$$e^{i(2\theta)} = \left(e^{i\theta}\right)^2 = (\cos\theta + i\sin\theta)^2 = (\cos^2\theta - \sin^2\theta) + i(2\cos\theta\sin\theta)$$

Equating real parts, we find

$$\cos 2\theta = \cos^2\theta - \sin^2\theta.$$

Solutions for Section E

1. (a) $f'(x) = 3x^2 + 6x + 3 = 3(x+1)^2$. Thus $f'(x) > 0$ everywhere except at $x = -1$, so it is increasing everywhere except perhaps at $x = -1$. The function is in fact increasing at $x = -1$ since $f(x) > f(-1)$ for $x > -1$, and $f(x) < f(-1)$ for $x < -1$.
   (b) The original equation can have at most one root, since it can only pass through the x-axis once if it never decreases. It must have one root, since $f(0) = -6$ and $f(1) = 1$.
   (c) The root is in the interval $[0, 1]$, since $f(0) < 0 < f(1)$.
   (d) Let $x_0 = 1$.

$$x_0 = 1$$
$$x_1 = 1 - \frac{f(1)}{f'(1)} = 1 - \frac{1}{12} = \frac{11}{12} \approx 0.917$$
$$x_2 = \frac{11}{12} - \frac{f\left(\frac{11}{12}\right)}{f'\left(\frac{11}{12}\right)} \approx 0.913$$
$$x_3 = 0.913 - \frac{f(0.913)}{f'(0.913)} \approx 0.913.$$

Since the digits repeat, they should be accurate. Thus $x \approx 0.913$.

5. Let $f(x) = \sin x - 1 + x$; we want to find all zeros of f, because $f(x) = 0$ implies $\sin x = 1 - x$.

Graphing $\sin x$ and $1 - x$ in Figure E.2, we see that $f(x)$ has one solution at $x \approx \frac{1}{2}$.

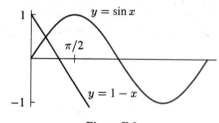

Figure E.2

Letting $x_0 = 0.5$, and using Newton's method, we have $f'(x) = \cos x + 1$, so that

$$x_1 = 0.5 - \frac{\sin(0.5) - 1 + 0.5}{\cos(0.5) + 1} \approx 0.511,$$

$$x_2 = 0.511 - \frac{\sin(0.511) - 1 + 0.511}{\cos(0.511) + 1} \approx 0.511.$$

Thus $\sin x = 1 - x$ has one solution at $x \approx 0.511$.

9. Let $f(x) = \ln x - \frac{1}{x}$, so $f'(x) = \frac{1}{x} + \frac{1}{x^2}$.
Now use Newton's method with an initial guess of $x_0 = 2$.

$$x_1 = 2 - \frac{\ln 2 - \frac{1}{2}}{\frac{1}{2} + \frac{1}{4}} \approx 1.7425,$$

$$x_2 \approx 1.763,$$

$$x_3 \approx 1.763.$$

Thus $x \approx 1.763$ is a solution. Since $f'(x) > 0$ for positive x, f is increasing: it must be the only solution.

Solutions for Section F

1. Between times $t = 0$ and $t = 1$, x goes at a constant rate from 0 to 1 and y goes at a constant rate from 1 to 0. So the particle moves in a straight line from $(0, 1)$ to $(1, 0)$. Similarly, between times $t = 1$ and $t = 2$, it goes in a straight line to $(0, -1)$, then to $(-1, 0)$, then back to $(0, 1)$. So it traces out the diamond shown in Figure F.3.

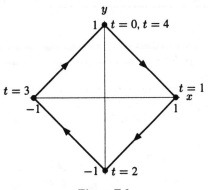

Figure F.3

5. The particle moves clockwise: For $0 \leq t \leq \frac{\pi}{2}$, we have $x = \cos t$ decreasing and $y = -\sin t$ decreasing. Similarly, for the time intervals $\frac{\pi}{2} \leq t \leq \pi, \pi \leq t \leq \frac{3\pi}{2}$, and $\frac{3\pi}{2} \leq t \leq 2\pi$, we see that the particle moves clockwise.

9. Let $f(t) = \ln t$. Then $f'(t) = \frac{1}{t}$. The particle is moving counterclockwise when $f'(t) > 0$, that is, when $t > 0$. Any other time, when $t \leq 0$, the position is not defined.

13. One possible answer is $x = -2, y = t$.

17. The ellipse $x^2/25 + y^2/49 = 1$ can be parameterized by $x = 5 \cos t, y = 7 \sin t, 0 \leq t \leq 2\pi$.

21. (a) C_1 has center at the origin and radius 5, so $a = b = 0, k = 5$ or -5.
    (b) C_2 has center at $(0, 5)$ and radius 5, so $a = 0, b = 5, k = 5$ or -5.
    (c) C_3 has center at $(10, -10)$, so $a = 10, b = -10$. The radius of C_3 is $\sqrt{10^2 + (-10)^2} = \sqrt{200}$, so $k = \sqrt{200}$ or $k = -\sqrt{200}$.

25. For $0 \leq t \leq 2\pi$

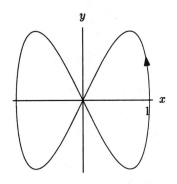

29.

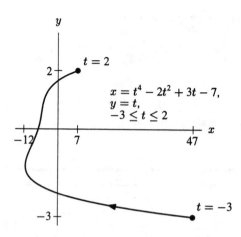

The particle starts moving to the left, reverses direction three times, then ends up moving to the right.

Solutions for Section G

1. (a) Eliminating t between

    $$x = 2 + t, \quad y = 4 + 3t$$

    gives

    $$y - 4 = 3(x - 2),$$
    $$y = 3x - 2.$$

    Eliminating t between

    $$x = 1 - 2t, \quad y = 1 - 6t$$

    gives

    $$y - 1 = 3(x - 1),$$
    $$y = 3x - 2.$$

    Since both parametric equations give rise to the same equation in x and y, they both parameterize the same line.
    (b) Slope $= 3$, y-intercept $= -2$.

5. We have $dx/dt = -2t \sin(t^2)$ and $dy/dt = 2t \cos(t^2)$. Therefore, the speed of the particle is given by

    $$
    \begin{aligned}
    v &= \sqrt{(-2t \sin(t^2))^2 + (2t \cos(t^2))^2} \\
    &= \sqrt{4t^2(\sin(t^2))^2 + 4t^2(\cos(t^2))^2} \\
    &= 2|t|\sqrt{\sin^2(t^2) + \cos^2(t^2)} \\
    &= 2|t|.
    \end{aligned}
    $$

    The particle comes to a complete stop when speed is 0, that is, if $2|t| = 0$, and so when $t = 0$.

9. The length is

    $$\int_1^2 \sqrt{(x'(t))^2 + (y'(t))^2 + (z'(t))^2} \, dt = \int_1^2 \sqrt{5^2 + 4^2 + (-1)^2} \, dt = \sqrt{42}.$$

    This is the length of a straight line from the point $(8, 5, 2)$ to $(13, 9, 1)$.

13. (a) The parametric equation describing Emily's motion is

$$x = 10 \cos\left(\frac{2\pi}{20}t\right) = 10 \cos\left(\frac{\pi}{10}t\right), \quad y = 10 \sin\left(\frac{2\pi}{20}t\right) = 10 \sin\left(\frac{\pi}{10}t\right) \quad z = \text{constant.}$$

Her velocity vector is

$$\vec{v} = \frac{dx}{dt}\vec{i} + \frac{dy}{dt}\vec{j} + \frac{dz}{dt}\vec{k} = -\pi \sin\left(\frac{\pi}{10}t\right)\vec{i} + \pi \cos\left(\frac{\pi}{10}t\right)\vec{j}.$$

Her speed is given by:

$$\|\vec{v}\| = \sqrt{\left(-\pi \sin\left(\frac{\pi}{10}t\right)\right)^2 + \left(\pi \cos\left(\frac{\pi}{10}t\right)\right)^2 + 0^2}$$

$$= \pi\sqrt{\sin^2\left(\frac{\pi}{10}t\right) + \cos^2\left(\frac{\pi}{10}t\right)}$$

$$= \pi\sqrt{1} = \pi \text{ m/sec,}$$

which is independent of time (as we expected). This is certainly the long way to solve this problem though, since we could have simply divided the circumference of the circle (20π) by the time taken for a single rotation (20 seconds) to arrive at the same answer.

(b) When Emily drops the ball, it initially has Emily's velocity vector, but it immediately begins accelerating in the z-direction due to the force of gravity. The motion of the ball will then be tangential to the merry-go-round, curving down to the ground. In order to find the tangential component of the ball's motion, we must know Emily's velocity at the moment she dropped the ball. Then we can integrate the velocity and obtain the position of the ball. Assuming Emily drops the ball at time $t = 0$, her position and velocity vector are

$$\vec{r}(0) = 10\vec{i} + 3\vec{k} \text{ and } \vec{v}(0) = \pi\vec{j}.$$

Thus, the ball has velocity only in the y-direction when it is dropped. In the z-direction, we have

$$\text{Acceleration} = \frac{d^2z}{dt^2} = -9.8 \text{ m/sec}^2.$$

Since the initial velocity 0 and initial height 3, we have

$$z = 3 - 4.9t^2.$$

The ball touches the ground when $z = 0$, that is, when $t = 0.78$ sec. In that time, the ball also travels $\pi(0.78) = 2.45$ meters in the y-direction. So, the final position is $(10, 2.45, 0)$. The distance between this point and $P = (10, 0, 0)$ is 2.45 meters.

(c) The distance of the ball from Emily when it hits the ground is found by finding Emily's position at $t = 0.78$ sec and using the distance formula. Emily's position when the ball hits the ground is $(10 \cos(0.078\pi), 10 \sin(0.078\pi), 3) = (9.70, 2.43, 3)$. The distance between this point and the point where the ball struck the ground is:

$$d \approx \sqrt{(10 - 9.70)^2 + (2.45 - 2.43)^2 + (0 - 3)^2} = 3.01 \text{ meters.}$$

Note that the merry-go-round doesn't rotate very much in the 0.78 sec needed for the ball to reach the ground, so our answer makes sense.